Science and
Other Cultures

Science and Other Cultures

Issues in Philosophies of
Science and Technology

Edited by

Robert Figueroa and
Sandra Harding

ROUTLEDGE
New York & London

Published in 2003 by
Routledge
29 West 35th Street
New York, NY 10001
www.routledge-ny.com

Published in Great Britain by
Routledge
11 New Fetter Lane
London EC4P 4EE
www.routledge.co.uk

Routledge is an imprint of the Taylor & Francis Group.
Printed in the United States of America on acid-free paper.

10 9 8 7 6 5 4 3 2 1

Cataloging in publication data is available from the Library of Congress.

ISBN 0-415-93991-7 (hb)
ISBN 0-415-93992-5 (pb)

Contents

Part III   *Tradition and Modernity: Issues in*
           Philosophies of Technological Change

Acknowledgments

This collection of papers originated in a National Science Foundation grant to the American Philosophy Association (APA) to develop a series of research activities focused on the topic of diversity issues in the philosophy of science. We, the editors, managed the grant for the APA for most of its term. The grant sponsored fourteen summer research projects and approximately thirty-six research presentations at four regional APA meetings from December 2000 through December 2001.

These APA programs were organized mostly by minority caucuses of the APA and APA-recognized philosophical societies including the Committee on the Status of Asian and Asian-American Philosophers and Philosophies, the Committee on the Status of American Indians in Philosophy, the American Indian Philosophy Association, the Committee on the Status of Women, the Society for Women in Philosophy, the Committee on Lesbian, Gay, Bisexual and Transgender People in the Profession, as well as the Committee on International Cooperation and the APA Board of Officers. This collection contains a selection of the essays developed through both processes.

We express our appreciation to the National Science Foundation, and especially to Program Director Rachelle Hollander, Ethics and Values Studies Program, without whose enthusiasm and assistance this project would not have existed. Richard Bett, Acting Executive Director of the APA, administered the grant for its first two years. The project greatly benefitted from his assistance and good advice. We also owe thanks to Elizabeth Radcliffe, Executive Director of the APA for the grant's final year. The grant project received valuable assistance in various stages from the APA Executive Board, Eric Hoffman, the former Executive Director of the APA in whose term the grant began, Julia Yearian and Linda Smallbrook in the APA National Office, and Bill Mann, Robin Smith, and Anita Silvers from the three regional APA divisions. Bill Lawson, of Michigan State University, managed the grant for its first year.

The APA appointed a splendid Advisory Board to oversee and assist the project. Joseph Rouse of Wesleyan University, Anita Silvers of California State University at San Francisco, and Elliot Sober of the University of Wisconsin responded quickly and effectively to our requests for proposal and manuscript reviews. Their thoughtful analyses improved the quality of the essays and of this collection. Stephanie J. Bird of Massachusetts Institute of Technology gave the project valuable assistance at a crucial point. We thank Damon Zucca, our editor at Taylor and Francis, for his enthusiasm for the collection and his good advice. Katherine Muir, a graduate student in the Department of Education at the University of California at Los Angeles, managed the manuscript review process. We owe Kate an immense debt for her skillful services. We also thank Tara Watkins, also a graduate student in the UCLA Department of Education, for her assistance in the final days of the project.

Not all editorial or authorial collaborations are pleasurable. Sandra thanks Rob for the very best one of the many I have had. Rob expresses his heartfelt gratitude and deepest appreciation for Sandra's kindness, wit, brilliance, and unswerving camaraderie.

Introduction

Sandra Harding and Robert Figueroa

Science and Values?

Science and values is a familiar topic for philosophers. Most standard philosophy of science readers have contained a small section of papers at the end of the volume under such a heading.[1] Yet, with but rare exceptions, those who wrote on such topics thought that they were addressing two distinct phenomena and a relation between them that science could control, at least in principle. "Science" was assumed to refer to inquiry processes that could be made value free and thus capable of transcending any particular cultural context. "Values" (or "society") was taken to refer to cultural and political values, interests, and ways of thinking that tended to sneak or march into scientific research processes and results, thereby restricting the growth of knowledge as well as often harming innocent people, or that shaped the uses of otherwise value-free sciences and their technologies. From this perspective, "Nazi science," Hiroshima, environmental destruction, the alienation of labor, and scientific sexism and racism were the consequence either of "bad science" or of the misuse and abuse of pure science's technologies and applications.

Thus, philosophy of science and the ethics of science could and should be addressed separately. So, too, could philosophy of science and philosophy of technology be addressed separately; the latter was not considered part of epistemology but rather of social philosophy. Moreover, the politics of science might be a topic for political scientists, but it was not a topic for philosophers since the "good science" that philosophers studied by definition had no politics; only "bad science" and the misuses and abuses of technologies and of applications of sciences could have politics. Finally, the productive character of cultural values and interests—the ways cultural beliefs and practices sometimes advance the growth of knowledge—could not be identified or examined as a philosophic topic.

Skepticism about the separability of science from society, its values and politics, began appearing in philosophy of science more than four decades ago. W. V. O. Quine (1960) proposed that scientific and everyday beliefs were linked in networks. How scientists theorized nature's order and chose to revise their hypotheses when faced with counterevidence depended in part on the ontologies, logics, and epistemologies they brought to their work, largely unconsciously, from their particular cultural contexts. Thomas Kuhn (1970) produced influential arguments claiming that to understand the history of scientific belief formation, one needed to focus not only on intellectual histories but also on the kinds of social histories of science that had begun to appear. Particular moments in the history of modern (Western) science had an "integrity" with their historic eras, he argued.

In subsequent decades such insights were expanded and refined through the work of philosophers, historians, and sociologists pursuing the legacy of Quine, Kuhn, and other historians, philosophers, and sociologists of that generation (Hess 1996). These scholars have shown in empirical detail and in principle how sciences and their cultures coconstitute each other; each cannot flourish or be understood apart from the resources provided by the other. Moreover, cultural elements permeate even the cognitive, technical core of Western sciences, whether this is conceptualized as scientific method, the metaphysics of modern sciences, formulations of the laws of nature, or some other component.[2] These scholars have shown how, at times, some cultural values and interests have been productive of knowledge—they have advanced scientific inquiry. Thus, values, interests, and culture should not be conceptualized only as blocking the growth of knowledge. To be sure, these social "constructivist" tendencies have met resistance within these disciplines and outside them; they are not yet uncontroversial. And a small minority of these scholars, mostly sociologists, have taken excessively constructivist positions, adopting a judgmental relativism from which their peers distance their work.

Yet, apart from such strengths and limitations of these studies, virtually all of them have remained contained by Eurocentric understandings of European sciences and technologies. The preoccupation with "high sciences" of European modernity has continued, with some exceptions, into the constructivist era; for the most part, physics and other formerly "pure" sciences have retained centrality as the unique model of good science. Partly for this

reason, these studies have rarely ventured to examine the empirical knowledge traditions of non-Western cultures. Nor have they set out to examine
critically Western sciences and their philosophies from the standpoint of
other cultures or their encounters with Western sciences. They have not
taken the standpoint toward knowledge production of exploited or marginalized groups within Western or other cultures such as women, racial
and ethnic minorities, disabled people, the poor, or lesbians, gays, and
transsexuals. Consequently, valuable as the post-Quinean and Kuhnian
studies are, they have not sought to gain a critical perspective on the assumptions characteristic of the "natives" of modern scientific cultures—
that is, of the professional, administrative, and managerial classes in the
West. Instead, they have distanced their work in this case from kinds of
self-critical objectivity-producing practices that they otherwise advocate.
Last but not least, for the most part, the post-Quinean and Kuhnian
philosophies of science and technology do not take overtly engaged positions; they retain political and ethical distance from the sciences-in-
cultures and cultures-in-sciences tendencies that they chart.

However, skepticism about the separability of science from social values, interests, and ways of thinking has also appeared in another school of
contemporary science studies, namely scholarly projects formulated in the
context of prodemocratic social movements that have arisen around the
globe.[3] Such projects have taken standpoints outside the conventional "exceptionalist" and "triumphalist" popular and scholarly understandings of
Western culture and its scientific and technological achievements (to
which we return below). From such positions, they have begun to examine
critically conventional philosophies of Western sciences and technologies,
conventional devaluations of other cultures' scientific and technological
traditions, and the history and recent practices of encounters between
knowledge systems of the West and its others.

The contributors to this collection bring a variety of concerns and perspectives to their studies of philosophic issues about social diversity and difference in particular scientific and technological contexts. Yet, all are concerned to situate more carefully both sciences and their philosophies in
their social environments. They do so in order to identify more clearly the
relevance of social context to standards of good science, good philosophy,
and the good society, and vice versa. Readers will detect a number of additional common themes in these essays, including some that take up or critically engage central concerns in mainstream philosophies of science and
technology. We will not here summarize the arguments of the papers; in-

stead, we identify a few of these themes that reappear throughout the collection. As we do so, we ask you to keep in mind that these essays are far more complex, interesting, and important than this brief overview can indicate.

Some Central Themes

Sciences in cultures; cultures in sciences.

The collection is organized around three such themes. The first section addresses philosophic issues about the complex ways in which sciences fit into their larger cultures and, in turn, harbor cultural values, interests, desires, and fears in even their most abstract elements.[4] As Westerners have always observed, in non-Western and minority cultures, ethical and epistemological elements cannot be separated, and that is what makes the knowledge projects of these cultures inferior to those of the modern West. However, these essays identify how it turns out that ethical and epistemological elements are also inseparable in Western scientific and technological practices.

In the opening essay, Robert Hood shows the inseparability of ethics and scientific standards in his examination of how the social context of a perceived health crisis can lead scientists to shift their standards for clinical trials from a concern to avoid false positives to a concern to avoid false negatives. His example is clinical trials in Africa on the effectiveness of AZT vaccines in protecting the fetuses of AIDS-infected pregnant women. In such contexts, scientists invoke ethical issues to justify their shift of standards. Yet, this does not turn the debate into one over ethics vs. science since ethical considerations also shaped selection of the traditional standards.

It proves easier to detect some kinds of ethical and other social elements of Western sciences from standpoints outside the latter—a point well understood in the ways traditional philosophies have valued the perspective of the "stranger" and his social distance from the worlds he observers. Alison Wylie's essay critically examines the persistently controversial feminist version of standpoint epistemology that develops a kind of "logic" or principle of how to bring an "outsider's" position to natural and social science research in a way more comprehensive and effective than conventional standards recommend. Three decades ago, standpoint analyses proposed a research methodology and epistemology that could distinguish characteristics of cultural values and interests that had positive effects on the growth of scientific knowledge from those that had negative effects. She disentangles what she sees as standpoint theory's essential core from its conflicted history and reframes it so that it complements philosophical science studies. Harding's essay shows how central insights of three decades of multicultural and postcolonial science and technology studies can expand the critical self-understanding of postpositivist philosophy of science that is evident, for example, in the "disunity of science" analyses. The perspectives from the new non-Western science studies—taking the standpoint of non-Western inquiry traditions—expand the scientific, epistemic, moral, and political significance of disunified sciences in ways that familiar postpositivist philosophies of science can welcome.

Hugh Lacey points out that from the standpoint of the roles that seeds play in poor peoples' agrarian cultures, we can more easily see that it is Western market values that make transgenic seeds seem universally valuable. Thus, we can't fully understand our own Western sciences and their philosophies if we stay within their conceptual frameworks. Lacey and James Maffie argue that many non-Western ways of developing empirical knowledge work just fine for the kinds of projects those cultures seek. The fact that their knowledge-systems are not ours is not in itself a good reason to reject the standards of their systems or the knowledge that such systems produce. If their kinds of goals were ours, we, too, would value their kinds of standards and knowledge. Furthermore, as Lacey and Maffie point out, non-Western ethical and cultural frameworks bring important benefits both to the sustainability of natural resources and the health of human communities. The goals of Western sciences are not the only desirable ones for peoples around the world or even, perhaps, the most desirable ones for Westerners.

When encounters between Western sciences and other cultures "go bad," it is important to avoid the temptation to conceptualize what went wrong as the consequence of the actions of (Western) "immoral madmen" conspiring to victimize already oppressed peoples. Instead, as Robert Crease points out, far more morally shocking is the fact that immoral consequences can follow from fully moral intentions. Crease examines disputes that have surrounded the actions of the U.S. doctors, politicians, and activists who sought to help the Marshall Islanders who had been exposed to fallout following a nuclear weapons test. We can, he writes, usefully understand this kind of encounter in terms of catastrophe "in the engineering sense of the word; what happens when a complex system grows out of synch with its environment, so it operates for a time in a domain of instability, until an incident causes the system to break down or operate in a drastically new mode. . . . The catastrophe is brought about not by the technical replacing the social—by science disabling critique and disarming social movements—but just the opposite; by it ceasing to inform practical action, ceasing to provide the basis for critique. That danger is especially strong in interactions with socially vulnerable populations and colonial environments." Western philosophies of science are part of modern Western world views more generally. The essays in this section suggest that philosophy of science, too, must be part of contemporary critical reevaluations of the Western world view. One might think this a silly suggestion if one assumes that philosophies of science no less than the sciences on which they reflect can be detached from cultural values and interests. Yet, the essays in this section provide strong evidence against such an assumption. Debates over standards for what is the best philosophic understanding of science also are issues about how to understand social relations.

Classifying people: science and technology at our service.

In the second section, contributors examine a number of ways in which Western sciences have been called upon to produce a politics of classification that, intended or not, turns out to serve the interests primarily of dominant groups at great cost to already vulnerable groups.[5] These essays show how scientific conceptions of physical and mental disability (Anita Silvers and Michael Ashley Stein, Licia Carlson), flexible cognition (Sara Waller), distinct races (Naomi Zack), and homosexuality (Margaret Cuonzo) serve discriminatory goals whether or not that was their intention. Silvers and Stein develop a proposal that promises to improve on existing efforts to temper the use of genetic information with justice. Their proposal reconceptualizes the classification of members of genetic minorities in ways that enable effective protective legislation for them. Carlson begins to examine the ways that critiques of normalcy developed in considerations of physical disability both are and are not useful in considering cognitive disability. Sara Goering points out how science and its medical technologies have been called upon to erase those signs of race and disability that the dominant culture devalues, increasing the impression that there is only one standard of beauty, and that that is the one that exclusively values the physical characteristics of dominant groups.

Naomi Zack shows how geographical grounds for racial distinctions are as unsupported by reliable evidence as are cultural, psychological, and biological grounds. Thus, the geographical arguments seem "to be no more than a rhetorical tradition deriving from Eurocentric reactions to contact with inhabitants of other places." Margaret Cuonzo proposes a new way of looking at vicious circularity in scientific studies. She takes as her example scientific ways of making animal homosexuality invisible or defining it as morally aberrant in order to argue for the unnaturalness and/or immorality of human sexual practices. She points out how a form of the "other minds" problem emerges in these studies, suggesting the necessity of far more tentativity in our conclusions about anyone's sexuality—animal or human—than sciences have generally practiced.

Technological change, tradition, and modernity.

In the last section, two papers focused on technological change raise questions about the relations between traditionalism and modernity that may well have implications for our understandings of scientific change, not only of technological change. Western sciences and technologies are conventionally characterized as modern, in contrast to the purportedly traditional knowledge systems that they are claimed to replace. "Modern" is supposed to represent Europe and its diasporas, progress, and the model of human "civilization." Yet, the essays by Andrew Feenberg and Junichi Murata contest such a contrast in considering several eras of technological change in Japan. Both introduce us to the insights of the founder of modern Japanese philosophy, Kitaro Nishida. Both show modernity and tradition in far more complex relations than can be understood through the conventional opposition between modernist universalism and traditionalist particularism.

Naturalizing philosophies of science and technology more extensively.

Four more themes can be traced through the arguments of these essays. These weave their topics into interests in contemporary philosophies of science that have not overtly featured issues of cultural diversity and difference. For example, a number of the essays argue for what one could call more effective "naturalization" of philosophy of science to actual scientific practices, in other and minority cultures as well as in modern Western sciences. Naturalized philosophies should replace residual idealizations of science that obscure the importance of considering the nature and effects of cultural difference on sciences and their philosophies. These essays argue that philosophies of science and technology that do not take account of how Western and non-Western sciences are in fact practiced in the West and around the globe tend to make faulty and parochial generalizations about what science is and can do.

Must philosophy of science be restricted to epistemology?

Must philosophy of science be restricted to concern with only how scientific beliefs are justified? Is philosophy of science thus only epistemology? Is there no possibility of any kind of logic of discovery? Is science fundamentally representations of nature or certain kinds of interactions with it? These interrelated questions have emerged in philosophies of science that have not been concerned with difference and diversity in the context of science—or at least, not so focused on it as are the contributors to this collection. (Williams 1991, Rouse 1987, 1996, Hacking 1983) Yet, these essays are concerned with contexts of discovery and how to make them accountable to the peoples such sciences study or otherwise affect. They are concerned with the patterned social processes that bring scientific problems, concepts, hypotheses, and standards for evidence to the starting point of scientific testing. They are concerned with the way empirical knowledge is developed, preserved, and revised through ethical and cultural practices—in the West as well as in non-Western cultures. Such accounts raise doubts about the value of restricting philosophy of science to epistemology.

Problems in the ethics of experimentation.

Cultures have always had encounters with each other that have scientific, technological, and political consequences. Yet, the ethics of experimentation on economically and politically vulnerable populations raises special issues and is the focus of concern in a number of these studies. (See especially the essays by Robert Crease and Robert Hood.) In some contexts such experimentation seems crucial to gaining the knowledge required to avoid future occurrences of the disasters such populations have experienced. Yet, the chances for cultural misunderstandings and for creating even further vulnerability are especially high when there is such a power imbalance between the experimenters and their subjects. Can ethical and scientific assumptions be separated in experimental contexts?

After exceptionalism and triumphalism.

Modern sciences and technologies have been claimed as the mark of the superior intellectual and cultural character of dominant groups and the mark of socially progressive cultures. Such an evaluation invokes both exceptionalist and triumphalist assumptions. Exceptionalism holds that only modern sciences could ever achieve such an exalted status and that they alone escape permeation by religious and cultural values that infest other cultures' knowledge systems. Triumphalism holds that the history of "real science" has been a history only of increasing achievements; any errors or bad effects have been the consequence not of real science but only of bad science or of politics. Indeed, "science" is usually used as both a descriptive and as an honorific term; it is taken to mark the observable achievement of identifiable standards, but also to mark them only when they are part of what already is counted as Western sciences, and only when the dominant culture is proud of them. Thus, Nazi science is not real science, but the identical eugenics practiced in Europe and the United States at the time was. (Proctor 1988) Phrenology—measuring intelligence by the shape of skulls—was thought to be science in its heyday, when it was developed and practiced by leading scientists of the day, but was later shown never to have been science at all once it had been abandoned. Such continual shifting between descriptive and honorific uses of the term makes defenses of exceptionalist and triumphalist Western science seem like just plain common sense. Skepticism about such practices, evident in many of these essays, helps us to get a more objective understanding of the history of our own and other cultures' knowledge systems, and of less romanticized ways of understanding the history and productive future of Western ways of thinking about nature and social relations.

Conclusion

Finally, we note that "culture," "science," "technology," and other terms that have been firmly defined in conventional philosophies of science and technology often are used in other ways in these essays. These authors want to think about them in philosophically fresh and more fruitful ways. Consequently, the familiar meanings and references of these terms are here often being contested. This can be unsettling to readers prepared to dismiss arguments that do not mean by "science" what, for example, the exceptionalist and triumphalist tradition has meant by it. Yet the familiar definitions of these terms are neither written in stone somewhere nor the result of empirical inquiry. "Science," in particular, has changed its meanings and referents again and again in the history of empirical inquiry. We recommend that readers simply be open to what new understandings they can gain when the formerly settled meanings and referents of such terms are expanded, shrunk, or otherwise shifted in order to expand our understanding of our philosophies and those of others and to enable more desirable philosophic, scientific, and social practices.

Notes

1. Or under "science and society."
2. See, for example, Hess 1995, Needham 1969, Restivo 1992, Schuster and Yeo 1986, Shapin 1994, Shapin and Shaffer 1985.
3. It has also appeared in antidemocratic social movements, such as religious and territorial fundamentalisms. (See Pels 1996) Some might think this sufficient reason to reject these social movement studies. However, after four decades of cultural studies of science and technology, it is no longer a plausible strategy simply to retreat to objectivist positions that ignore or deny the permeation of "good science" by culture and that ignore or deny the productive aspects of culture on scientific inquiry. See Castells 1997 for an illuminating account of the social conditions that produce similar protests against Liberal institutions (of which modern science is one such) from both the Right and the Left.
4. These issues are also pursued in essays in other sections.
5. Indeed, some critics would say that the most basic philosophic standards for "good science" in fact accomplish a similar classification that exalts "the West" at great cost to "the Rest": we produce good science; they do not.

References

Castells, Manuel. 1997. *The Power of Identity* (Vol. II, *The Information Age: Economy, Society and Culture*). Oxford: Blackwell Publishing Co.

Hacking, Ian. 1983. *Representing and Intervening*. Cambridge: Cambridge University Press.

Hess, David J. 1995. *Science and Technology in a Multicultural World: The Cultural Politics of Facts and Artifacts*. New York: Columbia University Press.

———. 1996. *Science and Technology Studies: An Introduction*. New York: Columbia University Press.

Kuhn, Thomas S. (1962) 1970. *The Structure of Scientific Revolutions*, 2nd ed. Chicago: University of Chicago Press.

Needham, Joseph. 1969. *The Grand Titration: Science and Society in East and West*. Toronto: University of Toronto Press.

Pels, Dick. 1996. "Strange Standpoints," *Telos* (108), 65–92.

Proctor, Robert. 1988. *Racial Hygiene: Medicine under the Nazis*. Cambridge, Mass.: Harvard University Press.

Quine, W. V. O. 1960. *Word and Object*. Cambridge, Mass.: MIT Press.

Restivo, Sal. 1992. *Mathematics in Society and History: Sociological Inquiries*. Dordrecht: Kluwer Publishing Co.

Rouse, Joseph. 1987. *Knowledge and Power: Toward a Political Philosophy of Science*. Ithaca, N.Y.: Cornell University Press.

———. 1996. *Engaging Science: How to Understand Its Practices Philosophically*. Ithaca, N.Y.: Cornell University Press.

Schuster, John A., and Richard R. Yeo, eds. 1986. *The Politics and Rhetoric of Scientific Method: Historical Studies*. Dordrecht: Reidel.

Shapin, Steven. 1994. *A Social History of Truth.* Chicago: University of Chicago Press.

——— and Simon Shaffer. 1985. *Leviathan and the Air Pump.* Princeton: Princeton University Press.

Smith, Dorothy. 1990. *The Conceptual Practices of Power: A Feminist Sociology of Knowledge.* Boston: Northeastern University Press.

Williams, Michael. 1991. *Unnatural Doubt: Epistemological Realism and the Basis of Skepticism.* Oxford: Basil Blackwell.

Sciences in Cultures, Cultures in Sciences

AIDS, Crisis, and Activist Science

Robert Hood

Introduction

HIV/AIDS activists, both in the United States and internationally, have charged that existing norms of research are unresponsive to the challenges of the global HIV/AIDS crisis.[1] Activists contend that HIV/AIDS requires a response significantly different from normal medical science and that HIV/AIDS research should be considered a crisis discipline. To explore how social phenomena shape the role of values in science, I focus on a recent controversy concerning international research about HIV/AIDS in developing countries.

Crisis Science

HIV/AIDS activists argue that in light of the global HIV/AIDS crisis, normal science should be replaced by an emerging view that might be called *crisis science*. Crisis science exists in relation to normal science as triage medicine to standard medical practice. The magnitude of HIV/AIDS, the time-dependency of developing treatments and prevention, and the consequences of failing to act lead activists to call for a radical approach to science. Crisis science involves a number of goals, among them the attempt to achieve results as quickly as possible, the right of patients to undertake greater risks in studies than would normally be tolerated, and the right for communities to participate in the direction of research. The activist position is that in a crisis, science should take greater risks because of the urgency of the situation.

The distinction between crisis science and normal science involves normative questions of how much of a chance a researcher should be willing to take, and what standards of proof should be demanded of researchers and for what purposes. Normal science can be characterized as cautious in the sense that it tries to minimize the likelihood that inferences are drawn based on chance alone. Since medical research considers samples of populations, by chance alone a researcher risks inferential errors from studying samples instead of the whole population. Statistical theory provides estimates of the probability of committing such errors by chance alone, and, normally, medical research employs cautious standards to try to decrease the probability that errors are due to chance. This cautious attitude helps increase the confidence that a positive result can be added to scientific knowledge that is not a result of random chance. Moreover, given the potential for grave errors, and given the complexity of medical research, using cautious standards generally serves to protect patients. Were researchers to take greater chances of new information being false by chance alone, the edifice of science would be less secure. Activists point out that a secure edifice of science, however, is not the only important social value at stake.

In particular, HIV/AIDS activists argue a cautious approach is not appropriate to the urgency of the HIV/AIDS crisis. The crisis of HIV/AIDS requires desperate measures, and there is not adequate time to establish the degree of confidence normally desirable. In addition to challenging standards of evidence, activists have also argued for—and paid for—alternate models of research. HIV/AIDS activists argue for rights to participate in drug trials and by so doing, undertake risks that are greater than those normally considered tolerable.[2] The issues concerning activist science are complex because they involve both ethical and scientific disagreements concerning how best to discharge obligations under conditions of uncertainty, and the discussion that follows moves between normative and scientific issues. I review a controversy that emerged over research into the prevention of mother-to-child transmission of HIV in developing countries to help diagnose and clarify these issues.

Controversy over Clinical Trials

Clinical trials were initiated in 1994 to address the problem that the standard of care in industrialized countries to prevent mother-to-child transmission of HIV is not affordable to the majority of people living in developing countries—to the majority of people in the world.[3] The purpose of the trial was to test a shorter, less expensive, and less intensive regimen of treatment in the hope that the short course would be safe and effective—and, most importantly, would be affordable to people in developing countries.[4] Starting in 1994, clinical trials involving over 12,000 pregnant women in developing countries were initiated by representatives of two international groups, the World Health Organization and UNAIDS (a United Nations agency coordinating international efforts to combat AIDS), and two agencies of the United States government, the National Institute of Health and the Centers for Disease Control and Prevention. In addition to this international support, the studies also had the support of local government officials in the study locations (Bloom 1998). The research protocols involved giving $80 or less of AZT, compared to the standard treatment, known as ACTG 076, which involves about $800 of AZT. In addition, the trials administered the treatment later in pregnancy for about one month, or only a fraction of the time required by the ACTG 076 standard of care in the United States.

Conflict emerged over the trial design—in particular, on the appropriate comparison group. Mothers in the trials in developing countries were given either the short course of AZT or a placebo—that is, the women and infants in the control group received no AZT. The fact that women from developing countries in the control group would get a placebo rather than the standard clinical treatment struck some as a straightforward ethical violation. Researchers stopped using placebo-controls in these particular studies a year after they began, when results emerged from a study in Thailand that showed the use of AZT in only the last four weeks of pregnancy could cut transmission rates by 50 percent (Marc Llemant and Vithayasai 1995; Strolberg 1998).

The majority of the literature on this controversy has focused on ethical issues, such as whether the there should be a single international standard for research subjects or whether global inequities in the allocation of health care resources justify placebo-controlled trials when an effective treatment exists but is unavailable in the host country (Schüklenk 1988; Angell 1997; Lurie and Wolfe 1997; Bloom 1998; Crouch and Arras 1998; del Rio 1998; Grady 1998; Lie 1998; Resnik 1998; Luna 1999; Schüklenk and Ashcroft 2000; Benatar 2001). In contrast, my focus is on how differing views of what is required under crisis conditions shape the practice of science. The two sides to the debate over international placebo trials reflect the distinction made above concerning crisis science: those in favor of placebo trials argued the placebo trials were necessary in order to show the degree to which the short course would be effective, whereas those opposed to the placebo trials seemed willing to forgo seeking additional evidence and argued that the dynamics of the HIV/AIDS crisis made this a special case on both scientific and ethical grounds.

Critics argued that placebo research in developing countries was not justified. For example, Public Citizens Health Research Group attacked these clinical trials on the grounds that randomized placebo-controls should not have been used when there was already an established standard of care for minimizing the transmission of HIV from mother to infant in industrialized countries, namely, the ACTG 076 protocol (Lurie and Wolfe 1997). The debate was heated, with one editorial claiming that the placebo trials are morally equivalent to the Tuskegee experiments (Angell 1997). The critics contend that as soon as the research community has good evidence that a treatment is more effective than a placebo, then researchers have an ethical obligation to discontinue use of placebos. They believe such evidence exists concerning ACTG 076 and, moreover, believe that research in industrialized countries concerning mother-to-infant HIV transmission applies in developing countries. Critics contend that studies that use placebos when effective treatment is available subordinate the welfare of human subjects to questionable research goals. Instead of placebo trials, they advocate comparative studies where subjects would get either the short-course treatment or the standard ACTG 076 protocol—that is, they advocate that the research in developing countries be conducted as it would be in industrialized countries.

Critics interpret international agreements governing conduct of clinical research as clearly proscribing placebo trials. The Helsinki Declaration, for example, grants subjects the right to receive "best proven diagnostic and therapeutic method" (World Medical Association 2001).[5] The *International Ethical Guidelines for Biomedical Research Involving Human Subjects* states that "researchers working in developing countries have an ethical responsibility to provide treatment that conforms to the standard of care in the sponsoring country, wherever possible" (Council for International Organizations of Medical Sciences 1993; see also Weijer and Anderson 2001). In addition, it is well established that for a trial to be ethical, a state of genuine uncertainty ("clinical equipoise") as to the comparative merits of the treatments under study must exist within the expert clinical community (Freedman 1987). Critics contended that there was not genuine uncertainty in this case and argued that they knew of no reason why study results could not be generalized across populations and across variations of HIV.

Critics also pointed out that for a trial to be ethical there must be a strong likelihood its results will be made available to the population in the host country. Since the per capita health expenditure in the region in question is often less than $10, critics charge that those conducting clinical trials knew the treatment would only in the most unlikely circumstances be made available to subjects in developing countries and so were unethical from the start, a point that has been discussed extensively (e.g., Grady 1998).

In addition to these ethical arguments, critics also contested the scientific merits of the trials. First, they argued that the performance of AZT in pregnant mothers is well understood (based on trials and clinical practice in industrialized countries). Those in developing countries are not, in their view, significantly different from those in industrialized countries. Furthermore, they contend that the use of placebo trials was unnecessary, for two reasons. First, an analysis of subgroup data from previous studies demonstrated that a short course was safe and effective (Lurie and Wolfe 1997). Second, they noted that a previous study in Thailand had already shown a shorter AZT regime to be safe and effective compared with the 076 protocol, and that, consequently, additional testing was unnecessary—especially when those getting placebos would not be getting any treatment. The Thailand study showed it was possible to reduce amounts of AZT while still achieving the efficacy of the ACGT 076 protocol. Notably, the Thailand study's directors refused to include a placebo group in the research because they thought using placebos would be unethical. Even without using placebos, the Thailand study showed the short-course AZT treatment resulted in a 50 percent reduction in mother-to-infant transmission (Marc Llemant and Vithayasai 1995; Lurie and Wolfe 1997). In sum, based on their analysis of the 076 subgroup data, and on the Thailand study, the critics concluded that sufficient evidence existed indicating the placebo trials were unnecessary.

Those defending the study also saw themselves as seeking the best possible scientific results as a foundation for making policy recommendations. In their view, placebo trials are the gold standard of clinical research and are most likely to yield the clearest results. Defenders respond to the moral arguments that, consistent with the general lack of health care in the developing countries of the study, the vast majority of the women in the study would not have received any medical treatment outside the study by arguing that participation in the placebo-controlled study did not deny them treatment they otherwise would have received, nor did it increase their risk of transmitting HIV to their infants. As Grady notes, "By participating in the study the women are not being *denied* treatment in the interests of science" (Grady 1998). Thus, the study did not deny access to anything because it did not otherwise exist. The justification for this turns on several distinctions, between clinical treatment and research on the one hand, and between misfortune and injustice on the other. Although it is unfortunate that people in developing countries cannot afford treatment, people who cannot afford treatment are not in this position because they were wronged; thus they have no right to health care (Crouch and Arras 1998).

Second, those defending the placebo study argue that placebo trials are necessary for scientific proof because the subject populations in developing countries are sufficiently different from people studied in the 076 protocol. Harold Varmus and David Satcher, for example, justified the trials on the grounds that the subject population of pregnant women in developing countries has a higher incidence of anemia, malnutrition, and various diseases, and "we do not have reliable data about safety and efficacy in populations from developing countries" (Varmus and Satcher 1997). The defenders do not address the concern that some patients in industrialized countries are also anemic and have multiple disorders. To the extent there is uncertainty about subjects in developing countries, then it is not clear why there wouldn't be uncertainty about similar patients in industrialized countries. Be that as it may, defenders point to other uncertainties—for example, the clades of the HIV virus in Europe and North America are different from those on the African continent and then again different from those in Asia. These differences in the virus could yield differences in drug behavior. Significantly, defenders of the trials were concerned that a short course of treatment might cause drug-resistant strains of HIV.

One defender of the placebo study, David Resnik, claims that placebo trials are necessary for science, arguing that "once scientists chose to tackle this research problem, empirical and methodological considerations support the case for using placebo-controls to meet the demands of scientific rigor" (Resnik 1998). Resnik and other defenders of placebo trials raise the concern that although comparative studies might have shown that the short-course treatment was less effective than ACTG 076, it would not have shown *how* much less effective or the degree to which it would still be worth implementing (Grady 1998). That is, a comparative study might not have had sufficient difference between results to show that a result was not due to chance. In general, if the difference between the two groups being compared is small, then it will take larger numbers of research subjects and could potentially take longer to show significant results. Thus, another reason defenders argue for placebo trials is because they will provide the clearest results faster than other study designs, particularly when the difference between groups is small. Interestingly, this attempt to answer the urgent need for results is consistent with what AIDS activists have asked for— namely, that science move quickly due to the urgency of the epidemic. Be this as it may, the most important point to make in response to this line of argument concerns the Thai study. Given that the Thai study demonstrated that a comparative study could show the degree of safety and efficacy with placebos, and given that other researchers dispute the need for placebo trials as essential to rigorous research, I am inclined to understand the remarks by those defending placebo trials as indicating a value preference for a high degree of confidence in the results of research. That is, they are concerned to know whether there might be unknown differences among the research population that might affect the safety or efficacy of the short-course treatment in developing countries.

Conclusion

The placebo studies discussed above would likely not have been allowed in an industrialized country such as the United States. The debate continues about the obligations of researchers concerning international research subjects, with a recent article lamenting that the debate has been reduced to a "war" characterized by the "effluvia of cynical political maneuvering" (Weijer and Anderson 2001). It is possible that this intensity of feeling stems from a deep recognition that, currently, none of the options concerning international research are adequate to the global HIV/AIDS pandemic.

I have been suggesting that there is significant disagreement among researchers concerning how best to discharge ethical obligations under crisis conditions. How should we cope with situations where the uncertainty that results from trying to avoid mistakes yields foreseeable harms that are of substantial magnitude and irreversible? As the discussion of the case above indicates, on one hand some are inclined to pursue placebo trials because of their view that this would yield the clearest results, even at the risk of appearing morally callous—because in their view, research is different from treatment, and does not require the same standards of care as treatment, and because of their contention that only placebo trials would yield clear results in a short time. Interestingly, this appears to be consistent with requests of HIV/AIDS activists for science to respond quickly to the urgent nature of the HIV/AIDS crisis. On the other hand, others are inclined to view the epidemic as a crisis requiring special consideration to protect, among other things, the rights of research subjects, particularly in the case of international subjects in developing countries.

The suggestion is that the source of the dispute can be traced to differences in views about how much certainty is enough and to different views about whether the preferred degree of certainty is achievable through comparative studies or only by placebo studies. I think that the situation would have been helped if the parties could have seen it as turning on values in the design of research studies—that both camps are concerned with rigor and both are concerned with ethics—but that their sense of the values internal to the practice of sciences leads them in different directions. For one group, the question concerns whether to focus on using placebo trials, which they see as providing clearer results; for others, it is whether to adopt comparative methods, which provide research subjects with at least the minimum standard of clinical care.

Notes

1. Approximately thirty-six million people are infected with HIV/AIDS in developing countries, with 25.3 million living in sub-Saharan Africa. About fourteen thousand people are infected with HIV every day, the majority of whom live in developing countries. Overwhelmingly, HIV/AIDS is concentrated in the world's poorest countries: 95 percent of all AIDS cases are in developing countries, and 89 percent of the world's HIV-infected population lives in the poorest 10 percent of countries. Approximately 5,500 people per day die from AIDS in sub-Saharan Africa. Approximately ten million African children have been orphaned by the epidemic. (UNAIDS 1999). The large-scale economic effects of the epidemic can be seen, for example, in South Africa, where economic growth is estimated to be reduced by 0.3–0.4 percent annually. By 2010 there will be an estimated 17 percent reduction in GDP compared to what it might have been without HIV/AIDS. (UNAIDS 2000). The situation is expected to continue to get worse. (Ainsworth and Over 1994; Matthews 1997; World Bank Policy Research Report 1999).

2. There are a number of differences between the current issues concerning research in developing countries and activist movements by gay men such as ACT-UP and Gay Men's Health Crisis in the early stages of the HIV/AIDS epidemic. Currently, there are standards of clinical practice, whereas, in the early days of HIV/AIDS, much less was known. Gay men and others at the time were asking for opportunities to participate in finding something, anything, that would work. Currently, a treatment is known and works, but is too expensive. Second, to a large extent the people in question in groups such as ACT-UP were extremely well informed about their care and about research, and so there was greater confidence in the ability to consent. Consent in developing countries is a more complex matter. (Crimp and Rolston 1987; Shilts 1987; Kinsella 1989; Crimp and Rolston 1990; Grmek 1990; Centers for Disease Control and Prevention 2001).

3. A number of strategies have emerged to address this problem. Pharmaceutical companies in industrialized countries have reduced prices on HIV therapies to help make developing-country access more realistic. Governments in developing countries have adopted compulsory licenses to allow cheaper production of HIV therapies. Researchers have looked for ways to develop HIV vaccines and therapies that are less expensive and easier to implement than existing treatments common in industrialized countries. Each of these strategies has pro-

voked significant controversy. For example, industrialized countries have re-
sisted lowering prices on HIV treatments, with some drug makers going so far
as court to block reductions. So far the reduction of prices by Western pharmaceutical companies has not significantly improved access to HIV therapies in
developing countries. The attempted use of compulsory licenses by governments in developing countries sparked conflict between such governments,
Western pharmaceutical companies, and the United States.

4. The issue of HIV transmission from mother to infant is a serious health problem in developing countries, where some 1,600 HIV-infected babies are born
every day. Estimates indicate mother-to-child HIV transmission accounts for 5
to 10 percent of infections worldwide, with higher rates of 20 percent or more
in regions with widespread heterosexual epidemics such as in sub-Saharan
Africa. (World Bank Policy Research Report 1999). In 1994 researchers demonstrated an effective way for minimizing mother-to-infant transmission of HIV,
known as AIDS Clinical Trials Group (ACTG) Protocol 076, which reduced
transmission rates from 25 percent to less than 8 percent. According to this
protocol, mothers are tested for HIV early in pregnancy, receive oral and intra-
venous doses of the antiviral drug zidovudine (AZT), and forego breastfeeding;
in addition, AZT is given to the infant for six weeks. In the United States and
other industrialized countries ACTG 076 has remained the standard of care.
Unfortunately, there are a number of problems in applying ACTG 076 to developing countries. Expense is one prohibiting factor: at the time of the trial it
cost more than $800 per mother, a sum that is several hundred times more than
what many developing countries are able to spend in an entire year per capita
on healthcare. It requires a substantial healthcare infrastructure and substantial compliance to be effective because it must be started early in pregnancy and
must be continued after birth. (Centers for Disease Control and Prevention
1994).

5. The Declaration of Helsinki has been redrafted since these trials so as to clarify
the conditions under which placebo trials may be used. Principle 29 now states:
"The benefits, risks, burdens, and effectiveness of a new method should be
tested against those of the best current prophylactic, diagnostic, and therapeutic methods. This does not exclude the use of placebo, or no treatment, in studies where no proven prophylactic, diagnostic, or therapeutic method exists."
This means that trial enrollment into studies may be offered only where there is
no consensus concerning treatment in the community of medical practitioners, and refers to genuine uncertainty about the best course of treatment.
(World Medical Association 2001).

References

Ainsworth, M., and M. Over. 1994. "AIDS and African Development." *The World
Bank Research Observer* 9(July): 203–40.

Angell, M. 1997. "The Ethics of Clinical Research in the Third World." *New England
Journal of Medicine* 337: 847–9.

Benatar, S. R. 2001. "Justice and Medical Research: a Global Perspective." *Bioethics*
15(4): 333–340.

Bloom, B. 1998. "The Highest Sustainable Standard: Ethical Issues in AIDS Vaccines." *Science* 279: 186–199.

Centers for Disease Control and Prevention. 1994. "Recommendations of the U. S. Public Health Service Task Force on the Use of Zidovudine to Reduce Prenatal Transmission of Human Immunodeficiency Virus." *MMWR Morbidity and Mortality Weekly Reports* 43: 1–20.

Centers for Disease Control and Prevention. 2001. 25 Notable HIV and AIDS Reports Published in the Morbidity and Mortality Weekly Report (MMWR), Morbidity and Mortality Weekly Report (MMWR).

Council for International Organizations of Medical Sciences. 1993. *International Ethical Guidelines for Biomedical Research Involving Human Subjects.* Geneva, CIOMS.

Crimp, D., and A. Rolston, eds. 1987. *AIDS: Cultural Analysis, Cultural Activism.* Cambridge: MIT Press.

Crimp, D., and A. Rolston. 1990. *AIDS Demo Graphics.* Seattle: Bay Press.

Crouch, R. A., and J. D. Arras. 1998. "AZT Trials and Tribulations." *Hastings Center Report* 28(6): 26–34.

del Rio, C. 1998. "Is Ethical Research Feasible in Developed and Developing Countries?" *Bioethics* 12(4): 328–30.

Freedman, B. 1987. "Equipoise and the Ethics of Clinical Research." *New England Journal of Medicine* 317(3): 141–5.

Grady, C. 1998. "Science in the Service of Healing." *Hastings Center Report* 28(6): 34–38.

Grmek, M. D. 1990. *History of AIDS: Emergence and Origin of a Modern Pandemic.* Princeton, N.J.: Princeton University Press.

Kinsella, J. 1989. *Covering the Plague: AIDS and the American Media.* New Brunswick, N.J.: Rutgers University Press.

Lie, R. K. 1998. "Ethics of Placebo-Controlled Trials in Developing Countries." *Bioethics* 12(4): 307–11.

Luna, F. 1999. "Corruption and Research." *Bioethics* 13(3–4): 262–271.

Lurie, P., and S. Wolfe. 1997. "Unethical Trials of Interventions to Reduce Perinatal Transmission of the Human Immunodeficiency Virus in Developing Countries." *New England Journal of Medicine* 337: 853–6.

Marc Llemant, M., and V. Vithayasai. 1995. *A Short ZDZ Course to Prevent Perinatal HIV in Thailand.* Boston: Harvard School of Public Health.

Matthews, R. 1997. "AIDS Epidemic Warning for South Africa." *Financial Times* (May 9): 4.

Resnik, D. B. 1998. "The Ethics of HIV Research in Developing Nations." *Bioethics* 12(4): 286–306.

Schüklenk, U. 1988. "Unethical Perinatal HIV Transmission Trials Establish Bad Precedent." *Bioethics* 12(4): 311–9.

Schüklenk, U., and R. Ashcroft. 2000. "International Research Ethics." *Bioethics* 14(2): 158–72.

Shilts, R. 1987. *And the Band Played On: Politics, People, and the AIDS Epidemic.*
    New York: St. Martin's Press.
Strolberg, S. 1998. Placebo Use Is Suspended in Overseas AIDS Trials. *New York
    Times.* Sec. A (February 19, 1998), 16.
UNAIDS. 1999. The Children Left Behind: UNICEF and UNAIDS Issue New Re-
    port on AIDS Orphans. UNAIDS.
UNAIDS. 2000. *Report on the Global HIV/AIDS Epidemic.* UNAIDS.
Varmus, H., and D. Satcher. 1997. "Ethical Complexities of Conducting Research in
    Developing Countries." *New England Journal of Medicine* 337: 1000–5.
Weijer, C., and J. Anderson. 2001. "The Ethics Wars: Disputes over International Re-
    search." *Hastings Center Report* 31(3): 18–20.
World Bank Policy Research Report. 1999. *Confronting AIDS: Public Priorities in a
    Global Epidemic.* New York: Oxford University Press.
World Medical Association. 2001. *Declaration of Helsinki.* Cedex, France: Ferney-
    Voltaire.

Why Standpoint Matters

Alison Wylie

Standpoint theory is an explicitly political as well as social epistemology. Its central and motivating insight is an inversion thesis: those who are subject to structures of domination that systematically marginalize and oppress them may, in fact, be epistemically privileged in some crucial respects. They may know different things, or know some things better than those who are comparatively privileged (socially, politically), by virtue of what they typically experience and how they understand their experience. Feminist standpoint theorists argue that gender is one dimension of social differentiation that may make such a difference epistemically. Their aim is both to understand how the systematic partiality of authoritative knowledge arises—specifically, its androcentrism and sexism—and to account for the constructive contributions made by those working from marginal standpoints (especially feminist standpoints) in countering this partiality.

In application to scientific knowledge, standpoint theory holds the promise of mediating between the extremes generated by protracted debate over the role of values in science. In this it converges on the interests of a good many philosophers of science who are committed to making sense of the deeply social nature of scientific inquiry without capitulating to the kind of constructivist critique that undercuts any normative claim to epistemic privilege or authority.[1] Moreover, it offers a framework for understanding how, far from compromising epistemic integrity, certain kinds of diversity (cultural, racial, gender) may significantly enrich scientific inquiry, a matter of urgent practical and political as well as philosophical concern. Despite this promise, feminist standpoint theory has been marginal to mainstream philosophical analyses of science—indeed, it has been marginal to science studies, generally—and it has had an uneasy reception among feminist theorists. My aim in this paper is to disentangle what I take to be the promising core of feminist standpoint theory from this conflicted history of debate and to formulate, in outline, a framework for standpoint analysis of scientific practice that complements some of the most exciting new developments in philosophical science studies.

Contention about Standpoints

Standpoint theory may rank as one of the most controversial theories to have been proposed and debated in the twenty-five-to-thirty-year history of second-wave feminist thinking about knowledge and science. Its advocates as much as its critics disagree vehemently about its parentage, its status as a theory and, crucially, its relevance to current feminist thinking about knowledge. In a special feature on standpoint theory published by *Signs*, Hekman describes standpoint theory as having enjoyed a brief period of influence in the mid-1980s but as having fallen so decisively from favor that, a decade later, it was largely dismissed as a "quaint relic of feminism's less sophisticated past."[2] On her account, standpoint theory was ripe for resuscitation by the late 1990s; it is now being reconstituted by new advocates, revisited by its original proponents, and in Hekman's case (one of the former), heralded as the harbinger of a new feminist paradigm.

Hekman's telling has been sharply contested by those aligned with now canonical examples of standpoint theorizing—Hartsock, Harding, Smith, and Collins, most immediately[3]—but on some dimensions the differences among her critics are as great as between any of them and Hekman. Some ask whether there is any such thing as "standpoint theory": perhaps it is a reification of Harding's field-defining epistemic categories, an unstable (hypothetical) position that mediates between feminist empiricism and oppositional postmodernism.[4] When specific positions and practices are identified as instances of standpoint theory, the question arises of whether it is really an *epistemic* theory rather than a close-to-the ground feminist methodology; to do social science as a standpoint feminist is to approach inquiry from the perspective of insiders rather than impose on them the external categories of professional social science, a managing bureaucracy, ruling elites.[5] Among those who understand standpoint theory to be a theory of knowledge, there is further disagreement about whether it is chiefly descriptive or normative, aimed at the justification of knowledge claims rather than an account of their production. And there is wide recognition that feminist standpoint theory of all these various kinds has undergone substantial change in the fifteen years it has been actively debated. As Hartsock observes, "standpoint theories must be recognized as essentially contested" ("Next Century," 93).[6]

As fractious as this recent debate has been, however, there are some things on which everyone agrees: whatever form standpoint theory takes, if it is to be viable it must not imply or assume two distinctive theses with which it is often associated:

> *First*, standpoint theory must not presuppose an *essentialist* definition of the social categories or collectivities in terms of which epistemically relevant standpoints are characterized.
>
> *Second*, it must not be aligned with a thesis of *automatic epistemic privilege*; standpoint theorists cannot claim that those who occupy particular standpoints (usually subdominant, oppressed, marginal standpoints) automatically know more, or know better, by virtue of their social, political location.

Feminist standpoint theory of the 1970s and 1980s is often assumed to be a theory about the epistemic properties of a distinctively gendered standpoint: that of women in general, or that defined by feminists who theorize the standpoint of women, where this gendered social location is a biological or psychoanalytic given, as close to an "indifferent" natural kind as a putatively social, "interactive" kind can be (to use Hacking's terminology).[7] The claim attributed to this "women's way of knowing" genre of feminist standpoint theory is that, by virtue of their gender identity, women (or those who critically interrogate this identity) have distinctive forms of knowledge that should be valorized.

It is not clear that anyone who has advocated standpoint theory as a theory of knowledge or research practice has endorsed either the essentialist or the automatic privilege thesis. Hartsock and Smith, for example, were appalled to find their explicitly Marxist arguments construed in essentialist terms (Hartsock, "Truth or Justice," *Standpoint Revisited*, 232; Smith 1997); the point of insisting that what we know is structured by the social and material conditions of our lives was to throw into relief the contingent, historical nature of what we count as knowledge and focus attention on the processes by which knowledge is produced. Hartsock is no doubt right that early arguments for standpoint theory have been consistently misread because many of the commentators lack grounding in Marxist theory.[8] I would extend this analysis. The systematic and, in this sense, the perverse nature of the misreadings to which Hartsock responds reflect exactly the thesis her critics deny; their social location (if not consciously articulated standpoint—a distinction to which I will return) seems to impose the limitations of categories derived from a dominant individualist ideology. Hartsock, Collins, Harding, and Smith all object to a recurrent tendency to reduce the notion of standpoint to the social location of individuals, a move that is inevitable, I suggest, if it is incomprehensible (to critics) that social structures, institutions, or systemically structured roles and relations could be robust enough to shape what epistemic agents can know.[9] On such assumptions, unless the standpoint-specific capacities of knowers are fixed by natural or quasi-natural forces (e.g., biogenetic or psychoanalytic processes), standpoints fragment into myriad individual perspectives, and standpoint theory reduces to the relativism of identity politics.

It has to be said that, in her rebuttals to Hekman and various other critics, Hartsock makes little mention of her early use of psychoanalytic theory (object relations theory) to account for how individuals internalize the power relations constitutive of a sexual division of labor (specifically, reproductive labor) and the associated gender roles.[10] If essentialism lurks anywhere, it is in this component of her original argument, and it is this that has drawn the sharpest criticism.[11] It was the use of object relations theory to develop feminist theories of science and knowledge that Harding challenged in 1986 when she argued that the epistemic orientation attributed to women could not be a stable or universal effect of psychoanalytic processes set in motion by interactions with female caregivers; the characteristics distinctive of women closely parallel those claimed by the advocates of a pan-African world view as typical for men as well as women (*The Science Question*, 167–179, 185). But her critique left standing the central and defining (Marxist) insights of standpoint theory as articulated by Hartsock.[12] Indeed, Harding drew attention to structural characteristics of the power relations that constitute marked categories in opposition to (as exclusions from) whatever is normative in a given context—the oppositions between colonial elites and those subject to colonial domination; between men and women/not-men—and she argued that these have powerful, if contingent, material consequences for the lives of those designated "other" in relation to dominant social groups. It is an empirical question exactly what historical processes created these hierarchically structured relations of inequality, and what material conditions, what sociopolitical structures and symbolic or psychological mechanisms, maintain them in the present. But these are precisely the kinds of robust forces of social differentiation that may well make a difference to what epistemic agents embedded in systemic relations of power are likely to experience and understand. The processes of infantile socialization described by object relations theory may play an important role but so, too, do the ongoing relations of production and reproduction—the different kinds of wage and sex-affective labor people do throughout their lives—that are at the center of Hartsock's epistemic theory and Smith's sociological practice.[13]

By the early 1990s a number of standpoint theorists and practitioners had explicitly argued that it is this historical and structural reading of standpoint theory that bears further examination; essentialist commitments, if they were ever embraced or immanent, were roundly repudiated.[14] In this case, the variants of standpoint theory that have been live options in the last decade need not be saddled be with a commitment to claims of automatic privilege. Like essentialist readings of standpoint theory, I suspect that attributions of automatic privilege persist not because anyone advocates them, but because they are necessary to counter deepseated anxieties about what follows if strong normative claims of epistemic authority cannot be sustained. Debates about the viability of standpoint theory often seem to be driven by the assumption that, unless standpoint theorists can provide grounds for a new foundationalism, now rendered in social terms, they risk losing any basis for assessing and justifying knowledge claims; unless standpoints provide special warrant for the knowledge produced by those who occupy them, standpoint theory devolves into a corrosive (now solipsistic) relativism.[15] Hekman protests that, although standpoint theorists routinely claim that "starting research from the reality of women's lives, preferably those who are also oppressed by race and class, will lead to a more objective account of social reality," in the end, these theorists "offer no argument as to why this is the case" ("Truth and Method," 355). Hekman is dissatisfied with Harding's appeal to the epistemic advantage of standpoints that produce less partial, less distorted, "less false" knowledge ("Truth and Method," 353–355; Harding 1991, 185–187), and she rejects out of hand Hartsock's references to standpoints that put us in a position to grasp underlying realities obscured by ideological distortion ("Truth and Method," 346; Hartsock "Historical Materialism," 299). Her objection seems to be that talk of better and worse knowledge can make no sense unless we have a firm grip on notions of truth and objectivity that are robust enough to anchor epistemic justification; standpoint theorists have invoked, but failed to deliver, epistemic foundations.

I believe there is another way of reading the claims central to standpoint theory. Nonfoundationalist, nonessentialist arguments can be given (and have been given) for attributing epistemic advantage to some social locations and standpoints, although they are not likely to be satisfying for those who hanker for the security of ahistorical, translocational foundations. But to get this reading off the ground, a number of key epistemic concepts need to be reframed, and a distinction central to standpoint theory needs reemphasis.

Situated Knowledge vs. Standpoint Theory

First, the distinction. A recurrent theme in responses to Hekman, among others, is an insistence that standpoint theory is concerned, not just with the epistemic effects of *social location,* but with both the effects and the emancipatory potential of *standpoints* that are struggled for, achieved, by epistemic agents who are critically aware of the conditions under which knowledge is produced and authorized.[16] Although the importance of standpoints in this second sense is emphasized in these exchanges, I believe that standpoint theorists should concern themselves with the epistemic effects of (systemically defined) social location as well as with fully formed standpoints.

On the first more minimal sense, the point of departure for standpoint analysis is commitment to some form of a *situated knowledge* thesis:[17] social location systematically shapes and limits what we know, including tacit, experiential knowledge as well as explicit understanding, what we take knowledge to be as well as specific epistemic content.[18] What counts as a "social location" is structurally defined. What individuals experience and understand is shaped by their location in a hierarchically structured system of power relations: by the material conditions of their lives, by the relations of production and reproduction that structure their social interactions, and by the conceptual resources they have to represent and interpret these relations.

Standpoint in the sense that particularly interests standpoint theorists is our differential capacity to develop the kind of a standpoint *on* knowledge production that is a "project" (Weeks, 101), a critical consciousness about the nature of our social location and the difference it makes epistemically. Standpoint theory is itself such a project, carried out both through the kinds of social research that take seriously the understanding of insiders—e.g., feminist research that starts from women's experience and women's lives (Smith 1990; Harding 1991)—and by feminist philosophers who are intent on creating a politically sophisticated, robustly social form of naturalized epistemology and philosophy of science. In either case, what is at stake is the jointly empirical and conceptual question of how power relations inflect knowledge: what systematic limitations are imposed by the social location of different classes or collectivities of knowers, and what potential they have for developing an understanding of this structured epistemic partiality.

On standpoint theory so conceived, it is necessarily an open question what features of location and/or standpoint are relevant to specific epistemic projects. For example, although any location or standpoint that "disappears gender" should be suspect,[19] we cannot assume that gender is uniquely or fundamentally important in structuring our understanding, or that a feminist standpoint will be the key to understanding the power dynamics that shape what we know. The project of developing critical consciousness—a jointly empirical, conceptual, and social-political enterprise—is the only way to answer questions about the epistemic relevance of a standpoint (in either sense) to specific epistemic projects.

But then the normative question reasserts itself: is there any basis for claiming that we should privilege the knowledge produced by those who occupy a particular location or standpoint? Does an analysis of the epistemic effects of social location or achieved standpoint provide a basis for justification or does it reinforce a social constructivism that ultimately gives rise to corrosive relativism? The inversion thesis that underpins most forms of feminist standpoint theory suggests that, when standpoint is taken into account, often the epistemic tables are turned. Those who are economically dispossessed, politically oppressed, socially marginalized and are therefore likely to be discredited as epistemic agents—e.g., as uneducated, uninformed, unreliable—may actually have a capacity, by virtue of their standpoint, to know things that those occupying privileged positions typically do not know, or are invested in not knowing (or, indeed, are invested in systematically ignoring and denying). It is this thesis that Hekman contests when she objects that no argument has been given for attributing greater objectivity to such standpoints.

Epistemic Advantage

The term "objectivity" (like truth) is so freighted it might be the better part of wisdom to abandon it. But for present purposes, I propose a reconstruction that may be useful in showing what a standpoint theorist can claim about epistemic privilege without embracing essentialism or an automatic privilege thesis.

As Hekman uses the term, objectivity is a property of knowledge claims. Objectivity is also standardly used to refer to conventionally desirable properties of epistemic agents: that they are neutral and dispassionate with regard to a particular subject of inquiry or research project. And sometimes it is used to refer to properties of the objects of knowledge.[20] Objective facts and objective reality are contrasted with ephemeral, subjective constructs; they constitute the "really real", as Lloyd puts it (1996), a broad category of things that exist and that have the properties they have independent of us; presumably Hacking's "indifferent" kinds are at the core of this category of objects of knowledge (1999, 104–106). As a property of knowledge claims, objectivity seems to designate a loosely defined family of epistemic virtues that we expect will be maximized, in some combination, by the claims we authorize as knowledge. Standard lists, from authors as diverse as Kuhn, Longino (1990), Dupré, and Ereshefsky, include, most prominently, a requirement of empirical adequacy that can be construed in at least two ways: as fidelity to a rich body of localized evidence (empirical depth), or as a capacity to "travel" (Haraway) such that the claims in question can be extended to a range of domains or applications (empirical breadth).[21] In addition, requirements of internal coherence, inferential robustness, and consistency with well-established collateral bodies of knowledge, as well as explanatory power and a number of other pragmatic and aesthetic virtues, may be taken as marks of objectivity collectively or individually.

Standpoint theory poses a challenge to any assumption that the neutrality of epistemic agents, objectivity in the second sense, is either a necessary or a sufficient condition for realizing objectivity in the first sense, in the knowledge claims they produce. Under some conditions, for some purposes, observer neutrality—disengagement, strategic affective distance from a subject—may be an advantage in learning crucial facts or grasping the causal dynamics necessary for understanding a subject. But at the same time, considerable epistemic advantage may accrue to those who approach inquiry from an interested standpoint, even a standpoint of overtly political engagement. The recent history of feminist contributions to the social and life sciences illustrates how such a standpoint may fruitfully raise standards of empirical adequacy for hitherto unexamined presuppositions, expand the range of hypotheses under consideration in ways that ultimately improve explanatory power, and open up new lines of inquiry.[22]

Likewise, there is no reason to assume that the qualities of empirical adequacy, consistency, explanatory probity, and the rest cannot be realized, in some combination, in the investigation of objects of knowledge that are not "really real," for example, in the study of social phenomena that are interactive. Certainly, objectivity in these cases may be sharply domain-limited; empirically adequate knowledge about an interactive social kind that transforms itself in the course of investigation will not travel very far, but it is no less objective for all that.

This last points to a key feature of the epistemic virtues that figure on any list of objectivity-making properties: they cannot be simultaneously maximized.[23] For example, the commitment to maximize empirical adequacy in understanding a rapidly transmuting interactive kind requires a trade-off of empirical depth against empirical breadth. Similarly, explanatory power often requires a compromise of localized empirical adequacy,[24] as does any form of idealization.[25] The interpretation of these requirements is open-ended; they are evolving standards of practice. The determination of how one virtue should be weighed against others is, likewise, a matter of ongoing negotiation, which can only be settled by reference to the requirements of a specific epistemic project or problem. None of the virtues I have identified as constitutive of objectivity in the first sense are context- or practice-independent; they are all virtues we maximize for specific purposes. That said, the list I cite consists of epistemic virtues that have proven useful in a very wide range of enterprises—virtually any in which success turns on understanding accurately and in detail what is actually the case in the world in which we act and interact.

If the objectivity Hekman has in mind were understood in this sense—as designating a family of epistemic virtues that should be maximized (in some combination) in the claims we authorize as knowledge—there would be no incongruity in claiming that contingently, with respect to particular epistemic projects, some social locations and standpoints confer epistemic advantage. In particular, some *standpoints* (as opposed to *locations*) have the especially salient advantage that they put the critically conscious knower in a position to grasp the effects of power relations on their own understanding and that of others.[26] The justification that an appeal to standpoint (or location) confers is, then, just that of a nuanced, well-grounded (naturalized) account of how reliable particular kinds of knowledge are likely to be, given the social conditions of their production;[27] it consists of an empirically grounded assessment of the limitations of particular kinds of knowers, of how likely they are to be partial, and how likely it is that the knowledge they produce will fail to maximize salient epistemic virtues.[28]

The Advantages of an Insider-Outsider Standpoint: A Framework for Analysis

Consider the kinds of epistemic advantage that may accrue to a particular type of standpoint invoked by quite diverse advocates of standpoint theory: that of a race, class, and gender disadvantaged "insider-outsider" who has no choice, given her social location, but to negotiate the world of the privileged, a knower who must understand accurately and in detail the tacit knowledge that constitutes a dominant, normative world view at the same time as she is grounded in a community whose marginal status generates a fundamentally different understanding of how the world works. Collins draws on the wisdom of black women domestics to illustrate what such an insider-outsider knows, and there are antecedents in the sociological literature as well as a number of parallel discussions in feminist contexts.[29] But one of the most compelling accounts of "what housecleaners know,"[30] one that affirms and extends Collins's central points, is a fictional account: Barbara Neely's murder mystery, *Blanche on the Lam*.[31] Blanche, on Neely's telling, clearly occupies a standpoint, not just a social location; she is sharp-tongued and incisive in her analysis of the conditions for survival that require her to know more, to know better, and to know more quickly, than those she works for. Consider the epistemic advantages of Blanche's standpoint that emerge with particular clarity in Neely's novel.

Blanche is a fill-in domestic for a rich white family in North Carolina whose help is on leave while they sojourn in their summer house. A murder has been committed, but you do not learn exactly who has died until late in the plot. So the story unfolds as Blanche learns the peculiarities and history and finally the murderous secrets of the family that temporarily employs her. At one point, she reflects on the "ass-kissing" behavior of a long-time family retainer: "If it's for real, it's pitiful" (*On the Lam*, 52), she says, but then observes that "a black man in America couldn't live to get that old by being a fool" (*On the Lam*, 60). Performing epistemic incompetence goes with the territory: "This is how we've survived in this country all this time, by knowing when to act like we believe what we've been told and when to act like we know what we know" (*On the Lam*, 73). In particular, conforming to expectations of epistemic *in*authority serves a purpose. As Collins observes, "Afro-American women have long been privy to some of the most intimate secrets of white society" (1991, 35),[32] at least in part because they are treated as epistemic incompetents. Neely, through Blanche, describes in detail how this works. Because Blanche is presumed stupid, and anyway of no account, she is largely invisible to the family she works for.[33] Time and again she gleans information that is critically important to her survival (literally and figuratively) from conversations conducted in her presence as if she were a piece of furniture, from messes she cleans up, garbage she disposes of, errands she is sent on: "As far as the Graces of the world were concerned, hired hands didn't think, weren't curious, or observant, or capable of drawing even the most obvious conclusions" (*On the Lam*, 185).

This asymmetry of recognition puts Blanche in the way of empirical evidence to which few members of the white community, not even the immediate family, would have access. But when puzzles arise that Blanche cannot resolve in terms of what she learns by observing the family directly, she mobilizes an extended network of other insider-outsiders whose experience has much in common with her's.[34] She contacts Miz Minnie:

> Because she knew the black community, Miz Minnie also had plenty of information about the white one. Blanche wondered if people who hired domestic help had any idea how much their employees learned about them while fixing their meals, making their beds, and emptying their trash. (*On the Lam*, 115)

She learns a wealth of detail about the history of the family she's working for. She learns about its money problems and domestic disputes, its jealousies, eccentricities, legal tangles, and, most important, its position in the white elite; who its members can count on as allies and where the lines of long-standing feuds have been drawn. This collateral knowledge is critical to Blanche's understanding of the situation in which she finds herself; it provides key resources for interpreting fragments of observed behavior as evidence of underlying motivation and encompassing social relations, for checking the robustness of local patterns she has already discerned, and testing the hypotheses she is forming to explain them.

But beyond gathering and cross-checking a wide range of empirical evidence, Blanche has much to say about the uses of evidence made possible (and necessary) by her standpoint that illustrates another dimension of the epistemic advantage that may accrue to insider-outsiders. At a number of junctures Blanche comments on the necessity for a woman in her position to develop a subtle and sophisticated set of inferential heuristics to do with the kinds of motivations that might inform the actions of her white employers. She details psychological profiles that characterize those who occupy positions of power and privilege, sometimes making clear how sharply they contrast with those that are typical for members of her own community.[35] "As a person whose living depended on her ability to read character" (*On the Lam*, 184), she clearly recognizes that this critical understanding is essential; she must be able to discern patterns in the behavior she observes, and to construct and assess explanatory hypotheses about the underlying causes of this behavior, at lightning speed and with unerring accuracy.

As Narayan develops this point, not only do the oppressed "have epistemic privilege when it comes to immediate knowledge of everyday life under oppression" (36), their experience fosters an inferential acuity with respect to the dynamics of oppression that those living lives of relative privilege do not have to develop. Insider-outsiders are alert to "all the details of the ways in which their oppression . . . affects the major and minor details of their social and psychic lives" (36); they grasp subtle manifestations of power dynamics and they make connections between the contexts in which these operate that the privileged have no reason to notice or, indeed, have good reason not to notice. In short, it is an advantage and a liability of subdominance that you may have to develop sharply honed skills of pattern detection and an expansive repertoire of robust explanatory models to survive as an insider-outsider.

It is important to recognize, however, that this epistemic advantage is neither automatic nor all encompassing. While an insider-outsider like Blanche may have particular advantage in understanding the dynamics of oppression close up, and may be especially likely to recognize the simultaneity of oppressions operating along multiple lines of difference,[36] a condition of oppression is very often unequal access to key epistemic resources: certain kinds of information; the analytic skills acquired through formal education; a range of theoretical and explanatory tools. Narayan observes that, because oppression is "partly constituted by the oppressed being denied access to education and hence to the means of theory production (which would include detailed knowledge of the history of their oppression, conceptual tools with which to analyze its mechanisms etc.)," it is to be expected that "the oppressed may not have a detailed causal/structural analysis of how their specific form of oppression originated, how it has been maintained and of all the systemic purposes it serves" (36). In short, recognizing that the oppressed have epistemic privilege in some areas "need not imply that [they] have a clearer or better knowledge of the causes of their oppression" (35–36). Factory workers in the Maquiladoras District will have intimate knowledge of how work disciplines are manipulated to extract maximum profit, but they may not have access to the background knowledge and information necessary to understand the international movement of capital responsible for bringing a factory to their district from West Virginia, or for moving it to a tariff-free trade zone in Indonesia or Thailand.

A final dimension of the epistemic advantage that accrues to Blanche, and any who use the resources of a location like hers to develop the political-epistemic *standpoint* of an insider-outsider, is the critical dissociation she has from the authoritative forms of knowledge that are born of and that serve (that legitimate, rationalize) positions of privilege. Blanche has no investment in maintaining the world view that her employers take for granted; she is suspicious of the presumptions of epistemic authority that underpin their confidence in what they think they know, and it is this that puts her in a position to outmaneuver them as they attempt to cover up the murder they have committed.[37] By virtue of having to know how the world looks from more than one point of view, an insider-outsider like Blanche has at hand a set of comparisons that throws into relief the assumptions that underpin, and confound, a dominant world view. As Collins describes the standpoint of an academic insider-outsider, the dissonance between what she knows as a black woman and what she has learned as a sociologist—the assumptions that "traditional sociologists see as normal"— throws into relief the situated nature and the partiality of what has typically been privileged as authoritative knowledge (1991, 49, 51).

What Collins draws attention to here is the capacity of standpoint theory to account for the contributions that insider-outsiders have made to various forms of systematic empirical research. Standpoint theory has the resources to explain how it is that, far from automatically compromising the knowledge produced by a research enterprise, objectivity may be substantially improved by certain kinds of nonneutrality on the part of practitioners. To extend the example cited earlier, it is the political commitment that feminists bring to diverse fields that motivates them to focus attention on lines of evidence others have not sought out or thought important; to discern patterns others have ignored; to question androcentric or sexist framework assumptions that have gone unnoticed and unchallenged; and, sometimes, to significantly reframe the research agenda of their discipline in light of different questions, or an expanded repertoire of explanatory hypotheses.

Some of these epistemic advantages may accrue to those who occupy the social location of insider-outsiders even if this does not incline them to develop critical self-consciousness about the epistemic implications of their social location. Consider the rapidly expanding body of research on the "archaeology of gender" that has taken shape in the last decade. It is largely due to women who have focused attention on a range of neglected questions about women and gender, but nearly half of those who attended the first "Archaeology of Gender" conference in 1989 disavow any affiliation with feminism.[38] While the dearth of contact with feminist literature in other fields has certainly limited the scope of their work, those working in the "gender genre"[39] have challenged androcentric and sexist assumptions in virtually all active fields of archaeological research, and they have successfully introduced questions about women and gender to the research agenda of the field as a whole.[40] My thesis is that the location of these practitioners as women in a strongly masculinist discipline has mitigated against their development of a feminist standpoint at the same time as it has created for them a decisive rupture, in the sociological sense. Their very presence in the field—specifically, their collective presence, as members of the first cohorts of archaeologists in which the representation of women exceeded 20 percent (Wylie 1997, 95–96)—disrupts the conventional assumptions about gender roles that underpin not only the institutionalized practice of archaeology but also its conceptual framework. This dissonance has sensitized some practitioners (mainly, but not only, women) to questions about gender inequality and gender ideology that were never considered so long as gender schemas remained unchallenged.[41] And in some cases, it has induced those working on questions about women and gender to develop a feminist standpoint.[42]

In short, *contra* Hekman, arguments have been given for ascribing *contingent* epistemic advantage to (some) subdominant standpoints. These are arguments that demonstrate that objectivity can sometimes be improved, and partiality reduced, when inquiry is approached from these standpoints, not in an abstract sense measured against an absolute, ahistorical, transcontextual standard, but with reference to one or another subset of the more homely virtues I have identified as constitutive of objectivity. When it comes to solving the complex puzzle posed in *Blanche on the Lam*, Blanche is a better knower than (most) members of the family she serves, the elite white community of which they are a part, and the authorities who investigate the murder, because she is in a position, by virtue of her social location and her insider-outsider standpoint, to get more and better evidence, to discern motivations more accurately, to make connections between causal factors more quickly, and to test and cross-test a wider range of explanatory hypotheses than virtually anyone else in Neely's story. Blanche's knowledge deserves to be treated as authoritative, with respect to the epistemic project she engages, because she maximizes empirical adequacy (of the localized-depth variety), establishes consistency with a wide range of collateral knowledge, and develops an explanatory account of particular critical probity.

Conclusion

Although Blanche's investigations are fictional and her epistemic project is local and pragmatic, the central points I have made about the salience of standpoints can be readily extended to research in the social sciences and well beyond. Wherever structures of social differentiation make a systematic difference to the kinds of work people do, the social relations they enter, their relative power in these relations, and their self-understanding, it may be relevant to ask what epistemic effects a (collectively defined) social location may have. And whenever commonalities of location and experience give rise to critical (oppositional) consciousness about the effects of social location, it may be possible to identify a distinctive standpoint to which strategic epistemic advantage accrues, particularly in grasping the partiality of a dominant way of thinking, bringing a new angle of vision to bear on old questions and raising new questions for empirical investigation.

Extended to philosophical science studies, standpoint theory complements the social naturalism and pragmatism evident in the proposals for reframing post-positivist philosophy of science suggested by an increasingly broad spectrum of philosophers of science. Advocates of standpoint theory in the sense outlined here are centrally concerned to understand science as a collective enterprise shaped by the kinds of factors identified by Solomon (2001). They share Longino's commitment to move beyond the rational-social dichotomy that has so deeply structured divergent traditions of science studies (2002), a commitment that, as Rouse and Hacking have argued, directs attention to the practice (rather than the products) of science as it unfolds in socially and politically structured fields of engagement.[43] And they share Kitcher's appreciation of both the need and the potential for reframing ideals of objectivity so that scientific success can be understood in explicitly normative, pragmatic terms (2001). Most important, standpoint theorists recognize that questions about what standpoints make an epistemic difference and what difference they make cannot be settled in the abstract, in advance; they require the second-order application of our best research tools to the business of knowledge production itself. And this is necessarily a problem-specific and open-ended process.

Acknowledgments

I thank Nancy Tuana for inviting me to participate in "Philosophical Explorations of Science, Technology and Diversity," a special session of the American Philosophical Association annual meetings (Eastern Division, December 2000) sponsored by the APA Committee on the Status of Women and funded by the National Science Foundation. I also thank an anonymous reviewer for the APA/NSF project and colleagues the University of Notre Dame and Wesleyan University for invaluable discussions of this paper and of feminist standpoint theory generally.

Notes

1. I have in mind four recent monographs that, in quite different ways, make this mediation their central objective: Joseph Rouse, *Engaging Science: How to Un-*

derstand Its Practices Philosophically (Ithaca, N.Y.: Cornell University Press, 1996); Helen Longino, *The Fate of Knowledge* (Princeton, N.J.: Princeton University Press, 2002); Philip Kitcher, *Science, Truth and Democracy* (Oxford: Oxford University Press, 2001); Miriam Solomon, *Social Empiricism* (Cambridge, Mass.: MIT Press, 2001).

2. Susan Hekman, "Truth and Method: Feminist Standpoint Theory Revisited," *Signs* 22(2) (1997): 341; hereafter cited in text as "Truth and Method."

3. Collins, Hartsock, Harding, and Smith all published responses that appeared with Hekman's article. Nancy C. M. Hartsock, "Comments On Hekman's 'Truth and Method': Truth or Justice?" *Signs* 22.2 (1997): 367–74; hereafter cited in text as "Truth or Justice." Patricia Hill Collins, "Comment on Hekman's 'Truth and Method': Where's the Power?" *Signs* 22(2) (1997): 375–81. Sandra Harding, "Comment on Hekman's 'Truth and Method': Whose Standpoint Needs the Regimes of Truth and Reality?" *Signs* 22(2) (1997): 382–91. Dorothy Smith, "Comments On Hekman's 'Truth and Method'," *Signs* 22(2) (1997): 392–8.

   In addition, in the same year, Sally J. Kenney and Helen Kinsella edited a special issue of *Women and Politics* 18(3) (1997) on feminist standpoint theory, subsequently published as *Politics and Feminist Standpoint Theories* (New York: The Haworth Press, 1997); hereafter cited in text as *Politics*. For an assessment of the debate generated by standpoint theory see especially Kenney, "Introduction," in Kenny and Kinsella (eds.), *Politics*, 1–6; Katherine Welton, "Nancy Hartsock's Standpoint Theory: From Content to 'Concrete Multiplicity'," in Kenny and Kinsella (eds.), *Politics*, 7–24; Nancy J. Hirschmann, "Feminist Standpoint as Postmodern Strategy," in Kenny and Kinsella (eds.), *Politics*, 73–92; and Hartsock's response, "Standpoint Theories for the Next Century," in Kenny and Kinsella (eds.), *Politics*, 93–102; hereafter cited in text as "Next Century." See also Nancy C. M. Hartsock, "The Feminist Standpoint Revisited," in *The Feminist Standpoint Revisited and Other Essays* (Boulder, Colo.: Westview Press, 1998), 227–48; hereafter cited in text as *Standpoint Revisited*. The focus of these discussions is Hartsock's early formulation of feminist standpoint theory: Nancy C. M. Hartsock, "The Feminist Standpoint: Developing the Ground for a Specifically Feminist Historical Materialism," in S. Harding and M. B. Hintikka, eds., *Discovering Reality: Feminist Perspectives On Epistemology, Metaphysics, Methodology and Philosophy of Science* (Boston: Reidel, 1983), 293–5; hereafter cited in the text as "Historical Materialism."

4. Sandra Harding, *The Science Question in Feminism* (Ithaca, N.Y.: Cornell University Press, 1986), 24–29; hereafter cited in text as *The Science Question*. See also Sandra Harding, *Whose Science? Whose Knowledge? Thinking From Women's Lives* (Ithaca, N.Y.: Cornell University Press, 1991), ch. 5.

5. Dorothy E. Smith, "Women's Perspective as a Radical Critique of Sociology," *Sociological Inquiry* 44 (1974): 7–14; "A Sociology for Women," in J. Sherman and E. T. Beck, eds., *The Prism of Sex: Essays in the Sociology of Knowledge* (Madison, Wisc.: University of Wisconsin Press, 1979), 137–87. Reprinted in Dorothy Smith, *The Conceptual Practices of Power: A Feminist Sociology of Knowledge* (Toronto: University of Toronto Press, 1990).

6. Welton notes an important shift in emphasis in Hartsock's own characterization of standpoint theory that involves a move "from outlining the substantive

content and difference of the feminist perspective, based upon the shared character of women's experience, to a more formal understanding of the functioning of a standpoint, without emphasis on the actual content of this perspective" (1997, 7). This point is also made by Hirschmann although she notes a
persistent emphasis in Hartsock's writing on the "notion of standpoint as a
methodology" (1997, 76) and on commonalities in "the process of developing a
standpoint," rather than in the content of the standpoints that emerge from
this process, that predates "Historical Materialism." Hartsock herself enumer-
ates a number of issues that have drawn critical attention and require further
analysis by standpoint theorists of "the next century": analysis of the status of
experience, especially the notion of collective experience; reassessment of the
factors (in addition to labor) that are constitutive of the experience distinctive
for different groups of people; development of a more detailed account of
"how experience becomes mediated and transformed into a standpoint"
("Next Century," 95).

7. Ian Hacking, *The Social Construction of What?* (Cambridge, Mass.: Harvard
   University Press, 1999), 100–124.
8. Hartsock makes this point in *Standpoint Revisited* (229, 233) with reference to
   the Marxist-derived account of standpoint theory that she presented in "Historical Materialism." See also Hirschmann's assessment of various ways in
   which critics of essentialist and universalizing tendencies are unfair to Hartsock's early formulation of standpoint theory (1997, 74–5).
9. See, in particular, Hartsock's discussion of this point. She objects that even
   sympathetic commentators continue to give the individual (individual perspectives, subjectivity) too much prominence in their formulations of standpoint theory and calls for a clearer recognition of "the importance of epistemological collectivity in the production of standpoint analyses" ("Next Century,"
   94).
10. In the Marxist-feminist analysis that Hartsock developed in "Historical Materi-
    alism," the role of psychoanalytic theory was to supply an account of the dis-
    tinctive content of a *feminist* standpoint that might be derived from the shared
    (gender-specific) experience of women. Object relations theory loses its cen-
    trality as Hartsock responds to critiques of these universalizing claims and
    moves away from a concern with content to the emphasis on similarities in the
    processes by which feminist standpoints take shape—the shift outlined by Wel-
    ton (1979). Welton, Hirschmann, and other contributors to Kenney and Kin-
    sella (1997) describe these processes as essentially social and political; com-
    monalities in experience become the basis for forming a collective identity and
    associated standpoint which, in turn, allows for the discursive constitution of
    experience as salient for understanding the world in standpoint-distinctive
    ways. Experience does not figure in this account as the autonomous foundation
    for a distinctive standpoint, but neither is it entirely a discursive construct as
    some postmodern critics have suggested. Hirschmann argues, in this connec-
    tion, that "while experience exists in discourse, discourse is not the totality of
    experience"; the possibility of reinterpreting experience, in the process of for-
    mulating a standpoint, suggests that "there must be something in experience
    that escapes, or is even prior to, language" (1997, 84). By extension, O'Leary ar-

gues that experience rather than identity should be treated as primary in the formation of a standpoint; it is the essentially interpretive process of articulating commonalities in experience that underpins the formation of collective identity; Catherine M. O'Leary, "Counteridentification or Counterhegemony? Transforming Feminist Standpoint Theory," in *Politics and Feminist Standpoint Theories*, ed. Sally J. Kenney and Helen Kinsella (New York: Haworth Press, 1997), 65.

11. There are intriguing parallels here with the use Keller made of object relations theory in her early discussions of the gendered character of scientific practice and with the hostile reactions she drew. Evelyn Fox Keller, "Gender and Science," *Psychoanalysis and Contemporary Thought* 1.3 (1978): 409–33; "A World of Difference," in *Reflections on Gender and Science* (New Haven: Yale University Press, 1985), 158–79. For an assessment sympathetic to Keller's project see Jane Roland Martin, "Science in a Different Style," *American Philosophical Quarterly* 25(2) (1988):129–40.

12. This is an argument I have made in more detail in a review essay, "The Philosophy of Ambivalence: Sandra Harding on 'The Science Question in Feminism,'" *Canadian Journal of Philosophy*, Supplementary Volume 13 (1987): 59–73.

13. See, for example, Smith 1990; Hartsock's discussion in "Historical Materialism" (286–90), and in "Next Century" (95).

14. For a review of these developments, see Helen E. Longino, "Feminist Standpoint Theory and the Problems of Knowledge," *Signs* 19(1) (1993): 201–12.

15. See, for example, O'Leary's discussion of the threat of relativism that arises from a "logic of fragmentation" that many have assumed to be inherent in standpoint theory (1997, 57); and Hirschmann's discussion of "universalist" critiques of standpoint theory (1997, 77).

16. This is a point Hartsock emphasizes in her earliest discussions of standpoint theory: "a standpoint is not simply an interested position (interpreted as bias) but is interested in the sense of being engaged" ("Historical Materialism," 285). She reaffirms this point in "New Century," where she emphasizes that the formation of a standpoint is a matter of developing an "oppositional consciousness . . . which takes nothing of the dominant culture as self-evidently true" (96–97) and in "Truth or Justice" where, quoting Weeks, she argues that "a standpoint is a project, not an inheritance; it is achieved, not given" (370). Kathi Weeks, "Subject for a Feminist Standpoint," in Saree Makdisis, Cesare Casarino, and Rebecca E. Karle, eds., *Marxism Beyond Marxism* (New York: Routledge, 1996), 89–118. This view of standpoints as a (collective) achievement is also central to the sympathetic commentaries assembled by Kenney and Kinsella (*Politics*). See especially O'Leary, Hirschmann, and Catherine Hundleby, "Where Standpoint Stands Now," in *Politics and Feminist Standpoint Theories*, ed. Sally J. Kenney and Helen Kinsella (New York: Haworth Press, 1997), 41.

17. Miriam Solomon offers an especially useful account of various ways in which such a thesis may be construed; "Situatedness and Specificity" (manuscript in possession of the author, 1997).

18. In discussions in which standpoint theory is treated as a resource for developing a response to normative issues (e.g., in feminist philosophy of law) or elab-

orating a "poststructuralist" research program (e.g., in communication or social work), standpoint in this first sense—as social location—is often emphasized: Amy Ihlan, "The 'Dilemma of Difference' and Feminist Standpoint Theory," *APA Newsletter on Feminism and Philosophy* 94(2) (1995): 58–63; Mary E. Swigonski, "Feminist Standpoint Theory and the Questions of Social Work Research," *Affilia* 8(2) (1993): 171–83; Julia T. Wood, "Gender and Moral Voice: Moving from Woman's Nature to Standpoint Epistemology," *Women's Studies in Communication* 15(1) (1992): 1–24. Here standpoints are characterized as gendered subject positions (Wood, 12); "a social position" from which "certain features of reality come into prominence and other aspects of reality are obscured . . . one can see some things more clearly than others" (Swigonski, 172); or a recognition that "knowledge is perspectival . . . necessarily shaped by . . . personal perspective [which is] in turn . . . shaped by the particulars of individuals' life experiences, their relationships with others, and their historical situations" (Ihlan, 59–60). See also Bat-Ami Bar On's characterization of standpoint theory as, in the first instance, a form of social perspectivalism, "gender is a constitutive element of experience" and "some perspectives are more revealing than others"; "Marginality and Epistemic Privilege," in Linda Alcoff and Elizabeth Potter, eds., *Feminist Epistemologies* (New York: Routledge, 1993), 83. And Sismondo's assessment that "feminist standpoint theory, and standpoint theory generally, makes the claim that there are social positions from which privileged perspectives on knowledge can be obtained"; "The Scientific Domains of Feminist Standpoints," *Perspectives on Science* 3(1) (1995): 49.

Respondents to Hekman object that such formulations obscure the power dynamics that constitute standpoints as a *collective* achievement, reducing them to the idiosyncratic perspectives of individuals and abandoning the political dimension of standpoint analysis. While an analysis of the epistemic effects of social location by no means exhausts what standpoint theory has to offer it does have valuable insights to offer, and it need not reduce to the apolitical appraisal of the limitations and capabilities of individual epistemic agents.

19. Helen E. Longino, "In Search of Feminist Epistemology," *Monist* 77 (1994): 481.

20. For an elaboration of these distinctions, see Elisabeth A. Lloyd, "Objectivity and the Double Standard for Feminist Epistemologies," *Synthese* 104 (1996): 351–81.

21. Helen E. Longino, *Science as Social Knowledge: Values and Objectivity in Scientific Inquiry* (Princeton, N. J.: Princeton University Press, 1990); Thomas S. Kuhn, "Objectivity, Values, and Theory Choice," in *The Essential Tension* (Chicago: University of Chicago Press, 1977); John Dupré, *The Disorder of Things: Metaphysical Foundations of the Disunify of Science* (Cambridge, Mass.: Harvard University Press, 1993). Marc Ereshefsky, "Critical Notice: John Dupré, *The Disorder of Things*," *Canadian Journal of Philosophy* 25(1) (1995): 143–58; Donna J. Haraway, "Situated Knowledges: The Science Question in Feminist and the Privilege of Partial Perspective," in *Simians, Cyborgs, and Women: The Reinvention of Nature* (New York: Routledge, 1991), 183–202.

22. This argument is made with reference to a number of research fields in two recent publications: Londa Schiebinger, *Has Feminism Changed Science?* (Cam-

bridge, Mass.: Harvard University Press, 1999); Angela N. H. Creager, Elizabeth Lunbeck, and Londa Schiebinger (eds.), *Science, Technology, Medicine: The Difference Feminism Has Made* (Chicago: University of Chicago Press, 2001).

23. This is a point Longino makes with respect to a related but different list of epistemic virtues (1994, 479). I have proposed a refinement and extension of Longino's list in "Doing Philosophy as a Feminist: Longino on the Search for a Feminist Epistemology," *Philosophical Topics* 23(2) (1995): 345–358.

24. The tension between explanatory power and empirical adequacy is especially clear when explanation is conceived in unificationist terms. Although Kitcher has significantly modified his position (2001), his response to worries about the trade-offs that his earlier account may require is instructive; Philip Kitcher, "Explanatory Unification and the Causal Structure of the World," in *Scientific Explanation*, Minnesota Studies in the Philosophy of Science, Volume XIII, ed. P. Kitcher and W. C. Salmon (Minneapolis: University of Minnesota Press, 1989), 410–508. I give an analysis of these tensions in "Unification and Convergence in Archaeological Explanation: The Agricultural 'Wave of Advance' and the Origins of Indo-European Languages," *The Southern Journal of Philosophy* 34 (1995): 1–30.

25. Nancy Cartwright, *How the Laws of Physics Lie* (Oxford: Oxford University Press, 1984); "Capacities and Abstractions," in *Scientific Explanation*, Minnesota Studies in the Philosophy of Science Volume XIII, ed. Philip Kitcher and Wesley C. Salmon (Minneapolis: University of Minnesota Press, 1989), 349–56; William C. Wimsatt, "False Models As Means to Truer Theories," in *Neutral Models in Biology*, ed. M. H. Nitecki and A. Hoffman (Oxford: Oxford University Press, 1987), 23–55. See also Kitcher's proposal of "significance graphs" that capture the evolving contextual interests responsible for shaping specific trade-offs between epistemic virtues such as generality, precision, and accuracy (2001, 78–80).

26. This is a point Hartsock emphasizes in response to Hekman and other recent critics. She observes that one key measure of epistemic advantage is the degree to which a particular standpoint puts one in a position to "grasp the interaction among the various determinants that constitute one's social location" (*Standpoint Revisited*, 237–8).

27. I take it that this is the form of epistemic advantage Harding claims for critically self-conscious standpoints under the rubric of "strong objectivity," *contra* Hekman's foundationalist interpretation (Harding 1991). See also Sandra Harding, "Rethinking Standpoint Epistemology: 'What Is Strong Objectivity'?," in Linda Alcoff and Elizabeth Potter, eds., *Feminist Epistemologies* (New York: Routledge, 1993), 49–82.

28. This proposal to treat claims to epistemic privilege as contingent and relative to independent epistemic virtues raises a question that has been debated since Harding characterized standpoint theory as an unstable mediation between feminist empiricism and feminist postmodernism: that of whether, on such a construal, standpoint theory does not collapse into a form of social empiricism (*The Science Question*, 136–62). Hundleby addresses this issue in response to reductive arguments presented by several prominent feminist empiricists; I endorse her recommendation that standpoint theory should be seen as comple-

mentary to sophisticated feminist empiricism rather than as a sharply distinct, competing position (1977, 25, 33).

29. Patricia Hill Collins, *Black Feminist Thought: Knowledge, Consciousness, and the Politics of Empowerment* (New York: Routledge, 1990); "Learning from the Outsider Within," in Mary Margaret Fonow and Judith A. Cook, eds., *Beyond Methodology: Feminist Scholarship as Lived Research* (Bloomington Ind.: Indiana University Press, 1991), 35–9.

    Prominent among the sociological antecedents to feminist discussions of standpoint theory is a deeply conflicted analysis of Merton's, which turns on a consideration of the epistemic advantages afforded sociologists by race diversity; Robert K. Merton, "Insiders and Outsiders: A Chapter in the Sociology of Knowledge," *American Journal of Sociology* 78(1) (1972): 13. Collins cites this discussion as well as Simmel's account of what sociological insights "strangers" may have to offer, and Mannheim's characterization of "marginal intellectuals" (1991, 36): Karl Mannheim, *Ideology and Utopia: An Introduction to the Sociology of Knowledge* (New York: Harcourt, Brace & Co., 1954 [1936]); George Simmel, "The Sociological Significance of the 'Stranger,'" in Robert E. Park and Ernest W. Burgess, eds., *Introduction to the Science of Sociology* (Chicago: University of Chicago Press), 322–27. For examples of how insider-outsiders may operate as researchers, see Freire's account of the research practice required to institute effective literacy programs and examples of participatory action research: Paolo Freire, *Pedagogy for the Oppressed* (New York: The Continuum Publishing Company, 1982 [1970]); Elizabeth McLean Petras and Douglas V. Porpora, "Participatory Research: Three Models and an Analysis," *The American Sociologist* 23(1) (1993):107–26.

    A number of feminists have discussed the epistemic implications of insider-outsider standpoints. In the analysis that follows I draw chiefly on Uma Narayan, "Working Together Across Difference: Some Considerations on Emotions and Political Practice," *Hypatia* 3(2) (1988): 31–48. See also the difference theorists discussed by O'Leary (1997) and Hartsock ("Next Century"), and Chela Sandoval, "U.S. Third World Feminism: The Theory and Method of Oppositional Consciousness in the Postmodern World," *Genders* 10 (1991): 1–24.

30. Louise Rafkin, "What Housecleaners Know," *UTNE Reader* (March–April 1995): 39–40.

31. Barbara Neely, *Blanche on the Lam* (New York: Penguin Books, 1992); hereafter cited in text as *On the Lam*. Other mysteries by Barbara Neely are relevant in this connection as well, especially *Blanche Passes Go* (New York: Penguin Books, 2001).

32. As Blanche puts this point: "a family couldn't have domestic help and secrets," (*On the Lam*, 95).

33. In *Blanche on the Lam* Neely draws a series of parallels between Blanche's invisibility and that which characterizes the experience of Mumsfield, a Down's syndrome adult who is a cousin of the main protagonists:

> He went on to mimic some of his fellow churchgoers, including the less than kind comments they made about others among them—comments made right in front of him, because his condition made him as invisible as her color and

profession made her. . . . All us invisibles are probably sensitive [about being presumed not to understand]. (103)

Neely revisits this point in *Blanche Passes Go*, circumscribing the significance of these common features of experience in a way that reinforces the points made by O'Leary, Hirschmann, and Hartsock about the complexity of the relationship between commonalities of experience and a consciously articulated, collective standpoint (1997).

> Because of his Down's syndrome, much of the world treated him the same way it treated her. So he knew what it meant to be invisible, to be assumed to be the dummy in the room, to be laughed at because of parts of himself over which he had no control. This gave them something in common. But she didn't think mutual mistreatment was a basis for friendship. (62)

34. It is an important feature of Neely's later novel, *Blanche Passes Go*, that a number of those she turns to in the black community—neighbors and acquaintances who work for the wealthy white families Blanche investigates—make it clear that they do not share her critical standpoint (87, 152).

35. For example, "Blanche had seen it so many times before it no longer amazed her—people too rich to worry about being fired from their jobs or evicted from their homes who seemed to seek the threat of total disaster that poor people sought to avoid" (*On the Lam*, 117).

36. An insider-outsider like Blanche may be less likely than race- and class-privileged feminists, for example, to assume that any one dimension of difference is fundamental or essential. This point is discussed at some length by Collins (1991) and is eloquently argued by the Combahee River Collective, "A Black Feminist Statement," in Alison M. Jaggar and Paula S. Rothenberg, eds., *Feminist Frameworks* (Englewood Cliffs, N.J.: McGraw-Hill, 1984), 202–9.

37. As Hartsock observes, this critical distance underpins a complex analysis: "It is worth remembering that the vision of the ruling groups structures the material relations in which all parties are forced to participate and, therefore, cannot be dismissed as simply false," (*"Next Century,"* 96).

38. Alison Wylie, "The Engendering of Archaeology: Refiguring Feminist Science Studies," *Osiris* 12 (1997): 80–99.

39. This is the term Conkey and Gero use to refer to this growing tradition of non-feminist archaeological research on questions about women and gender (1997); Margaret W. Conkey and Joan M. Gero, "Gender and Feminism in Archaeology," *Annual Review of Anthropology* 26 (1997): 411–37.

40. I develop this argument in more detail in "Doing Social Science as a Feminist: The Engendering of Archaeology," in *Science, Technology, Medicine: The Difference Feminism Has Made*, ed. Angela N. H. Creager, Elizabeth Lunbeck, and Londa Schiebinger (Chicago: University of Chicago Press, 2001), 23–45.

41. I use the term gender schemas in the sense elaborated by Virginia Valian, *Why So Slow? The Advancement of Women* (Cambridge, Mass.: MIT Press, 1999).

42. This has had the effect not only of opening up a range of new lines of research but of mobilizing interest in a number of practical and political questions about how the gender structures evident in archaeology are created and main-

tained. A particular focus for these discussions has been the organization of archaeological labor in various contexts, employment and reward structures, and typical patterns of recruitment and training in archaeology. See, for example, contributions to the section "Gender and Practice" in *Gender and Archaeology*, ed. Rita P. Wright (Philadelphia: University of Pennsylvania Press, 1996), 199–280; and contributions to Margaret C. Nelson, Sarah M. Nelson, and Alison Wylie (eds.), *Equity Issues for Women in Archaeology*, Archaeological Papers of the American Anthropological Association, No. 5 (Washington, D.C.: American Anthropological Association, 1994).

43. For example, Ian Hacking, "The Self-Vindication of the Laboratory Sciences," in *Science as Practice and Culture*, ed. Andrew Pickering (Chicago: University of Chicago Press, 1992), 29–64.

A World of Sciences

Sandra Harding

The Information Society and Indigenous Knowledge Movements

During the last four decades, expanded globalization has been accompanied by increased interests in cultural identity. Indeed, social theorists see the latter as an expectable response to the increased homogenization and universalization of cultural forms created by the globalization of economies, politics, and of culture itself.

Both tendencies have generated distinctive interests in scientific and technological knowledge and practices. On the one hand, the base of the global economy has shifted from industrial manufacture to the production and management of information. Witness the way money speeds around the world with the tap of a computer key, leaving changed patterns of labor and leisure in its wake, bringing down some governments and stabilizing others, and making some groups immensely wealthy (for sometimes only a very short time) and others yet further impoverished. It becomes more and more difficult to see Western modern sciences as only Western; they do indeed seem to be truly international. Yet at the same time, paradoxically, it has also become more difficult to see their direction as independent of powerful social and political interests and desires.

On the other hand, a global movement has begun to establish intellectual property rights for indigenous knowledge; indigenous knowledge ministries have been established in at least several countries; the United Nations and other organizations have sponsored indigenous knowledge conferences around the world; and the World Wide Web permits an ongoing global networking and exchange of information about the past, present, and possible futures of indigenous knowledge systems. (For example,

49

see *The Indigenous Knowledge and Development Monitor.*) Moreover, during these four decades, there has emerged what is by now a flourishing literature, supported by conferences and other scholarly practices, about multicultural and postcolonial science and technology studies. At the same time as interests in preserving and developing local knowledge systems increase, international networks and increasingly widely shared assumptions about nature embedded in such systems also give them a kind of global status.

How is contemporary European-American philosophy of science positioned in relation to these social processes in which new interests in sciences and technologies play such active roles? Are these relevant contexts for philosophic thought? What changes could or should occur in Western philosophies of science as a consequence of such reflection? Here I focus on just one possible beginning of an enlightened response to such questions from the North, namely consideration of the already existing links between postpositivist philosophies of science and main philosophic themes in multicultural and postcolonial science and technology studies. Some sort of historical destiny has led these two philosophic movements to the brink of more extensive and fruitful engagement, and this essay tries to map the paths to that possible juncture.

The next section notes some already existing valuable links between the two scholarly fields. Section 3 briefly identifies several conceptual shifts in postpositivist philosophies of science that make additional important openings to issues raised in the other science studies. Section 4 notes five residual unity assumptions in the Northern philosophies that work against a more robust engagement with multicultural and postcolonial science studies. Section 5 shows how themes emerging from four recent multicultural and postcolonial science studies literatures help to overcome the residual unity assumptions so as to encourage a more robust engagement between the two movements.

Valuable Links

We can already identify some valuable resources for linking these two movements. For one, a few authors who have been alert to more general multicultural and postcolonial struggles in both the North and the South have already begun to reflect on what could be appropriate philosophies of science and technology for prodemocratic thinkers (e.g., Braidotti et al., 1994; Haraway 1989, 1991, 1997; Maffie 2001; Turnbull 1993; Verran 2001; Watson-Verran and Turnbull 1995).[1] Secondly, philosophic reflection on issues raised by this other science studies movement has appeared in the writings of scholars from around the world who work in science studies fields other than philosophy, especially in history, anthropology, and science education (e.g., Cobern 2001; Hess 1995; Selin 1997).

In the third place, Northern feminist epistemologies and philosophies of science have largely focused on European-American sciences, yet they have been concerned to change how science is practiced, and to do so in ways that expand social justice and equality. Like the multicultural and postcolonial science studies, and in contrast (for the most part) to the rest of philosophy and sociology of science, they have been prodemocratically politically engaged. Such commitments to such engagement create potentially powerful links between philosophies of science, North and South. Fourth, and relatedly, Northern feminists, again like multicultural and postcolonial studies, have been concerned to show how some prodemocratic political and cultural commitments can advance the growth of knowledge: politics and culture can be productive of knowledge, not just inevitably destructive of it as conventional philosophies have held. Northern science studies has recognized this phenomenon, but has done little to explore possibly systematic effects of different kinds of politics and culture on the growth of knowledge.

Additionally, certain familiar assumptions about the positive role of "difference," diversity, and disunity in conventional philosophies of science can be articulated in ways conducive to dialogue. For example, testing hypotheses across different observers and groups of observers has always been recognized as a valuable way to enhance the growth of knowledge through enabling the identification of merely local values and interests that have shaped such hypotheses and their tests. What if the goal were not to eliminate all such values and interests, but instead (as indicated above), to evaluate whether there are patterns of productive relations between culture and knowledge, and between prodemocratic culture and knowledge that can advance democratic social relations as well as the growth of knowledge? Sixth, scientific methods have valued multiple kinds of interactions with nature's order—different research methods. How scientists interact with nature both enables and limits what they can know about it. So if different cultures characteristically interact with their environments in different ways, each should be a repository of valuable different empirical knowledge. What would a philosophy of science look like that could engage with this value of "difference"?

Finally, it must be noted that the histories of science upon which philosophers draw have not completely ignored the importance to Northern modern scientific and technological change of information, insights, and technologies produced in non-Western cultures. Mainstream philosophers of science certainly are aware of the achievements of ancient Greek and Egyptian cultures, later Arabic and Islamic cultures, and Chinese and Indian cultures, and of their contributions to the emergence and development of modern Western sciences (Lach 1977; Needham 1954, 1969; Sabra 1976). The problem is not complete ignorance but rather that Westerners have always had trouble figuring out how to conceptualize such achievements. Are they scientific elements of prescientific knowledge systems? Prescientific elements that become scientific only when integrated into Western sciences? Purely speculative or purely technical and, thus, not scientific at all? What is needed is a better philosophy of Western modern science that can engage with such historical facts in empirically and theoretically reasonable ways.

Most significantly for my project here, main tendencies in philosophy of science have made conceptual moves during this period that make philosophy fertile ground for a more robust engagement with philosophic issues raised by the multicultural and postcolonial literatures. Yet central parts of the field still harbor kinds of unity of science assumptions and preoccupations that tend to block interest in such an engagement, though these, too, are currently receiving increasing skeptical attention. What would a philosophy of science look like that could grasp the complex relations between unifying and disunifying tendencies in the very best scientific research? I turn first to the conceptual shifts and then to the damaging residues of the unity thesis.

Useful Conceptual Shifts

For the last four decades, main tendencies in philosophy of science have critically focused on assumptions that there is fundamentally one and only one coherent set of empirical claims that constitutes real, or ideal, science. That is, widespread scepticism has arisen about the unity-of-science thesis. This holds that there is one world, one "truth" about it, and one and only one science that can, in principle, accurately represent that "truth." Criticism of such a thesis has important implications for the project of multicultural and postcolonial science studies, since if the unity thesis turned out to remain plausible in any of its standard forms, the ideal of a world of sciences would not. Note also that the thesis covertly requires that there be one and only one kind of ideal knower—presumedly the "rational man" of Enlightenment Liberal political philosophy.[2]

In different ways, Quine (1960), Kuhn (1970), and Feyerabend (1975) opened the way for rejecting the ideal, though all three worked entirely within the framework of Western modern sciences. They showed how important the possibility of lack of singularity and even of "harmonious integration" in belief could be to the advance of scientific knowledge.[3] These critics argued that inevitable continuities between everyday and scientific belief ensured that the best scientific beliefs at any time would represent only one possible set of reliable beliefs. Moments in the history of science exhibited an "integrity" with their historic era, as Kuhn put the point. And sciences' conceptual shifts, which reorganized most, but not all, existing data into new kinds of comprehensible patterns, were as important to the growth of knowledge as was the steady work of gathering evidence for or against favored hypotheses and of solving puzzles within prevailing conceptual frameworks. It was always possible to save existing hypotheses in the face of apparently counter evidence through various ad hoc justifications. Yet, the growth of science importantly depended upon comparing the adequacy of any given hypothesis with as many alternatives as possible, as Feyerabend insisted.

By the mid 1990s, historians and philosophers of science could overtly claim to disown not only the ideal of a single science that could, in some sense or other, in principle, contain all scientific knowledge, but also the presumption that the unity of science advocates had, with but rare exceptions, ever desired or even imagined such a possibility (Cartwright 1995; Dupre 1993; Galison and Stump 1996; Wylie 1999). It was only the harmonious integration of the sciences with each other that they intended to promote, not the singularity of science, as Hacking (1996) convincingly argues. Thus, one might conclude that at least the singularity ideal has finally been put to rest. This four-decade history of philosophy of science and science historiography seems to recognize what could reasonably be called a "world of sciences"—at least apart from the issues multicultural and postcolonial science and technology studies raise, which are largely invisible to these postpositivists.

However, the disunity analyses remain on the frontier of philosophy of science, not yet fully accepted or explored in the field. Moreover, the unity ideal remains the standard view of science in the media, as well as among many intellectuals in other fields. For example, the so-called science warriors focused in large part on the purported cognitive relativism of feminist and cultural studies of science. Such relativism was the consequence, as they saw it, of these studies' criticisms of the kinds of purportedly value-neutral objectivity and rationality ideals characteristic of philosophies of science. While some of these unity defenders expressed puzzlement at the criticisms of the neutrality standards for objectivity, rationality, method, and the like, others seemed enraged by them. (See, for example, Gross and Levitt 1994). The unity ideal seems to be held by many people not simply as a probably or even possibly reasonable proposal, but rather as a central component of their personal, professional, and public identities. It appears to be held as a moral belief that is used to distinguish themselves and their chosen community from others, including, especially, women and Europe's others.

In all fairness, it is not easy to develop a philosophy of science that avoids both assuming the unity ideal and falling into or adopting damaging forms of relativism. As has often been noted, epistemological relativism and absolutism are two sides of the same coin. So, the threat of a damaging relativism drives people back into the absolutes of the unity ideal in spite of plausible arguments against the latter. Unity idealists generate a "relativism panic" when faced with criticisms that the standards of modern science are too weak, as has been argued by feminists, the multicultural and postcolonial science studies, and elements in postpositivist philosophy of science. However, we must note, too, that the unity ideal can have progressive scientific and political effects in cultures where resistance to traditional intellectual and political pariochialisms is an ongoing struggle. And to give up the unity ideal is not to give up valuing unifying strategies of connecting different fields, models, theories, methods, concepts, and other elements of sciences. So, there are several reasons why it seems to many not possible, or perhaps even desirable, to fully abandon all of the insights that were captured by the problematic unity ideal.

I shall use the term "unity ideal" here to refer to both the singularity and "harmonious integration" theses, in contrast to the kinds of unifying strategies indicated above. It is the insistence on the ideal of harmony and consensus in postpositivist philosophies, unbalanced by equally high regard for disharmony and conflict in the production of knowledge, that I propose is problematic and that serves to direct philosophic eyes away from the multiplicity of functional inquiry processes and their consequent diverse knowledge systems around the world. At least in the West, the subterranean network of concepts that enable problematic features of the unity ideal to continue to flourish must be uprooted if the ideal is firmly to be abandoned. Elements of such a project have been underway for some time. Yet they have not been linked so as to encourage exploration of philosophic implications of other cultures' viable scientific traditions. I do not think this is an accident. It may be relatively cost-free these days to disavow the ideal of a singularity of science when the latter is described as reductionist. But it can be quite another matter to give it up when it is described as the assumption that modern Western science alone has the most desirable resources with which to grasp nature's order.[4] It is difficult for Westerners to countenance that others might, in some respects, in both practice and in principle, have produced better resources for understanding certain aspects of nature in the future as well as in the past. My point is that it is not clear that the ideal of a science unified at least by its distinctively Western (and androcentric) assumptions and integration into standard Western histories of "The West" and its others actually has ceased to provide the framework within which most philosophy of science occurs. In such respects the ideal has seemed to survive its reputed demise.

At least five nodes of the problematic network of belief supporting the unity of science thesis can be identified. They are by no means completely independent of each other, but rather contribute to constituting each other's apparent plausibility. (No doubt there are additional such sites in postpositivist networks of belief that could have been the focus of the discussion here.)

Residual Unity-Supporting Assumptions

Internalist vs. externalist epistemology.

The first is the assumption that one must hold only an "internalist," or only an "externalist," epistemology of science, grounded in a similar such history of science. If one makes such an assumption, it appears impossible to imagine that the cognitive, technical core of modern science—whether one defines that in terms of scientific method, the formally expressed laws of science, or some other criterion—could be both accurate about nature's order and also culturally mediated. As indicated earlier, Quine and Kuhn can be regarded among those who started off rethinking this dichotomy. An alternative conceptual practice examines how sciences and their societies coconstitute each other as each tries to manage effectively desired interactions with its environment. Many postpositivist philosophers of science engage in just this practice, yet most rarely consider how such overtly political social projects as European expansion, the control of women's labor, the expansion of bourgeois entitlements, or the commitment to a Liberal political philosophy have coconstituted the cognitive technical core of particular modern Western sciences.

No logic of discovery.

A second residual assumption rejects the idea of a "logic of discovery." This assumption restricts philosophy of science to epistemology—the justification of scientific belief—and restricts scientific method to the testing of hypotheses (Williams 1991; Rouse 1996). Some four decades ago N. R. Hanson (1958) tried unsuccessfully to establish such a logic. There were so many compelling arguments raised against such a project that there has been no overt attempt to take it up since (Caws 1967). Of course postpositivist philosophers commonly note that the research directions sciences take respond to questions of interest to their cultures, yet there is little attempt to reflect on the systematic scientific and epistemic limits to knowledge, and to the standards for "good science," created by the degrees of access different groups have to getting their questions to the starting point of research projects. However, feminist science studies and, especially, the uses of standpoint theory in these studies, can be understood to have traced a kind of logic of discovery in other terms (Harding 2003), as also has postcolonial science studies.

Eurocentric exceptionalism.

In the third place, it seems fair to say that recent philosophy of science, like most other fields, assumes that modern Western sciences are uniquely universally valid. This appears to be a residue of the "exceptionalist" claim that modern European American societies alone of all the world's cultures have managed to produce transcultural, universally valid claims of many sorts, of which scientific claims provide just one example. Ethics, literary and visual aesthetics, and political philosophy are other areas where Western beliefs have been claimed to be universally valid. New histories and ethnographies of modern sciences have already deeply challenged this claim. The demise of the exceptionalist assumption makes possible recognition that the agents or subjects of scientific knowledge are not only, or even primarily, individuals, but rather distinctive cultures with their particular knowledge projects. It encourages scepticism about the purported reality and desirability of the assumption that "word and object" can be cleanly distinguished (e.g., Verran 2001).

Representationism.

The fourth residue of the unity thesis is "representationism." (Hacking 1983, Rouse 1996). Is it best to conceptualize science as fundamentally a set of sentences—ideally, formal expressions of the laws of nature? Or is science more accurately and usefully conceptualized as effective systematic interactions with the world? No doubt it is foolish to assume that science has any kind of essence. Having said that, it seems valuable to focus on the priority of exemplars of scientific practice in Kuhn's work, in Foucault's discussions of the way social sciences developed out of—on the model of—the institutions of the prison and clinic, and in Hacking's discussion of the neglected importance of scientific intervention. These are three influential arguments skeptical of representationism. To advocate an alternative "interactionist" view of science is to undermine the possibility of a damaging relativism in a "world of sciences," a point to which I return below.

Eurocentric triumphalism.

Finally, the appeal of the unity thesis persists past its purported demise because of "triumphalist" assumptions that modern Western sciences and their rationality exemplify the height of human achievement. Most of the aspects of social relations of which Westerners are proudest are said to be the consequence of modern sciences and their forms of rationality. On this account, unexpected or destructive effects of modern sciences are attributed to politics, culture, or something else said to be outside science proper. Triumphalist assumptions have been countered in many ways, most directly through "dystopian" arguments about modern Northern sciences' causal links to an impressive array of disasters, including escalating environmental destruction, the Holocaust, Hiroshima, de-development and maldevelopment in the Third World, and perpetual services to militarism and to class exploitation's rationalization of labor—not to mention to other forms of racism, sexism, and the subjugation of other cultures. Triumphalism produces a distinctly Eurocentric system of accounting for science's intended and unintended nature and effects.

In order to distance even postpositivist philosophies of science more fully from these assumptions that support the unity of science thesis, we need an even more objective perspective on the latter.[5] Here is where the other science studies can be especially valuable.

Multicultural and Postcolonial Science and Technology Studies

This is by now a flourishing field.[6] Here I can only offer a brief overview of a few of the main concerns in these accounts that bear on the conceptual practices supporting problematic residues of the unity of science thesis. I do so by identifying four distinctive intellectual and political movements that have organized the field.

Comparative ethnoscience studies.[7]

These studies examine and reevaluate the achievements of non-Western systematic empirical inquiry, and they look also in methodologically symmetrical ways at Western inquiry processes.[8] Historians and ethnographers have produced most of these studies. They point out that it is not just modern Western sciences that have systematically developed the kinds of knowledge that enable cultures to survive and flourish in their natural and social environments; so, too, have non-Western sciences. Nor is it only non-Western sciences that embed knowledge about themselves and their environments in religious, cultural, and other local belief systems and practices; so, too, do Western sciences (as the postpositivist studies also argued). Moreover, it is not just modern Western sciences that are dynamic, continually developing better ways to interact with their part of nature's changing order and borrowing ideas and techniques from other cultures. All cultures' sciences do so as they encounter changing environments and ideas and technologies that interest them. And all cultures' knowledge systems retain beliefs that no longer fit well with the faster-changing strands of their networks of belief. Thus, these comparative ethnoscience studies show how all knowledge systems, including modern sciences, contain at least traces of their particular histories and ongoing practices; they are all "local knowledge systems" in this respect.

Elsewhere, I have argued that we should expect all empirical beliefs to exhibit such inevitable and, in important respects, desirable cultural features if we reflect on four characteristics of why and how cultures engage in inquiry processes (Harding 1998). Different cultures occupy different locations in nature's heterogeneous order, and each has interests in asking questions about its environment. These two features alone will tend to give them different bodies of systematic knowledge and of the systematic ignorance that inevitably accompanies any pattern of knowledge. But the inevitability that such bodies of knowledge will not neatly fit together to complete the "big jigsaw puzzle," mapping nature's order (one figure of the representation of nature that the singularity [unity] of science thesis generated), becomes even clearer in light of two further considerations. Different cultures bring different discursive resources to their queries about their environments—metaphors, models, and narratives that make sense within the distinctive religious, national, and other kinds of projects of their cultures. For example, environmental movements have brought to scientific inquiry models of earth as a spaceship and as a lifeboat. These models conflict with the Christian models of God's nature forever provisioning His children and with earlier models of nature as a inexhaustible cornucopia of resources. Yet, each way of thinking about nature drew attention to important aspects of nature's order.

Fourth, cultures tend to organize scientific inquiry in the ways that they organize any other kind of social labor. How one interacts with nature both enables and limits what one can know about it, as Western scientists have always argued in defending the value of scientific method. Finally, all four conditions of inquiry processes—locations in nature, interests, discourses, and ways of organizing inquiry—are shaped by a culture's (or subculture's) "location" in social relations, as standpoint theorists have pointed out (Harding 2003). Peasants and aristocrats, men and women, the rich and the poor, gain access to different parts of nature, have often oppositional interests, develop different discourses, and organize inquiry differently through the activities characteristically assigned to those in their social positions.

Of course it is no news to Northern science and technology studies that Northern sciences, too, are always shaped by local cultural projects and accessible natural resources. Yet, the problematic residues of the unity-of-science thesis identified earlier can look more starkly improbable from the perspective of these symmetrical studies of Northern and other cultures' inquiry systems.

An interesting development here is the possibility of comparative studies of particular philosophic notions. Several studies have focused on, for example, culturally distinctive concepts of scientific method and of truth in the history of Western sciences (e.g., Schuster and Yeo 1986; Shapin 1994). Now similar such projects are beginning to appear which focus on comparing Western and non-Western belief systems. For example, Watson-Verran and Turnbull (1995) have looked at different cultures' "universalizing practices" (or unifying practices), which enable the transportation of information about nature through the generations in cases where writing is not available. James Maffie (2001) has edited a collection of papers that examines the nature and role of truth claims or their nearest equivalent in different cultures' philosophies. Selin (1997) provides rich materials for further such analyses. Such accounts have the effect of reducing the impression of the exceptionalism of Western sciences while enabling richer understandings of inquiry practices in the North and elsewhere.

Science and empire studies.

A second focus of these other science studies has been on whether and how the rise of modern sciences in Europe and the "voyages of discovery" are causally related. Such studies emerged from the work of historians, economists, and political theorists.[9] Is it entirely the irrelevant coincidence suggested by conventional European histories that the voyages of discovery and the "birth of modern science" started in Europe during the same period?[10] To the contrary, each required the success of the other for its own successes. And this symbiotic relation between European expansion and the advance of modern sciences continues today, according to critics of the North's Third World development policies.

These accounts chart how the voyages of discovery produced exactly the information about nature's order needed for Europeans successfully to establish global trade routes and settlements in the Americas and, eventually, Australia, New Zealand, Africa, and elsewhere around the globe. The success of such expansionist projects required advances in navigation, cartography, oceanography, climatology, botany, agricultural sciences, geology, medicine, pharmacology, weaponry, and other scientific fields that could provide information enabling Europeans to travel far beyond the boundaries of Europe and to survive encounters with unfamiliar oceans, lands, climates, flora, fauna, and peoples. In turn, the production of such information required European expansion. Expansion enabled Europeans to forage in other cultures' knowledge systems, absorbing into European science useful information about the new environments Europeans encountered, new research methods, and new conceptual frameworks. Expansion permitted Europeans to appropriate access to nature around the globe, so to compare, contrast, and combine together new observations of nature's regularities. (Consider, for example, how important such access was to Darwin.) Moreover, through expansion, potentially sophisticated competitors to European science were vanquished (intentionally or accidently—for example, through the effects of European-introduced infectious diseases) along with the flourishing of such cultures, and sometimes their very existence. Thus the voyages of discovery and subsequent European expansionist projects greatly contributed to the way modern sciences flourished specifically in Europe, rather than also, or instead, in other cultures of the day.

The argument here is not that all European sciences equally benefitted from European expansion. They did not; for example, modern physics was largely developed during a somewhat similar process of increased travel, warfare, and social migration that occurred within Europe. Nor is the argument that Europeans, including scientists, were vicious, evil-minded creatures who always intended the destructive consequences of their expansionist projects. Indeed, European expansion was justified through appeals to improving the quality of life for Europeans and bringing civilization to the savages, doing God's work in the world, and in other "noble" ways not so very different from how the North's Third World development policies have been justified since their inception in the 1950s. The issue is not evil intent, though such can sometimes be identified, but rather the unpredicted or, to the perpetrators, unimportant consequences of well-intended projects. It is this disinterest in predicting the bad consequences for others of scientific projects, and the evaluation of such consequences when they are encountered as unimportant, that should be far more disturbing than the occasional appearance of overtly evil intent: good science and good intentions can have horrible ethical, political, and scientific consequences.

These studies highlight important ways in which modern Northern sciences and their cultures coconstituted each other. And the third focus in these other science studies prevents us from thinking that such Northern colonial and imperial projects are entirely a matter of history.

Development science and technology policy.

The North's five decades of Third World development policies have been justified by humanitarian appeals to bring the so-called undeveloped societies up to the standard of living of the developed Western societies. This was from the beginning conceptualized as achievable only through the transfer to the undeveloped societies of Northern scientific and technological expertise, rationality, and the democratic political forms that these purportedly encourage (e.g., see Snow 1964). Yet the policies responding to such appeals have, intentionally or not, largely continued the earlier imperial and colonial pattern of directing the flow of natural, human, and other economic resources from the South to the North. Maldevelopment and dedevelopment for the majority of the world's peoples have tended to be the now-obvious effects of the introduction of scientifically rational agriculture, manufacturing, health care, and so forth into the already economically and politically disadvantaged societies of the Third World. Simultaneously, it is the "investing classes" in the North and their allies in the South who have in fact been "developed" by these policies as, by now, even the mainstream Northern press regularly reports (Bass 1990; Braidotti et al. 1984; Sachs 1992; Shiva 1989). The flow of money, natural resources, "brain power" (as in "brain drains"), other kinds of labor power and social benefits have continued to flow from South to North—in the last half century no less than in the earlier 450 years. It is the North that overreproduces; the South cannot reproduce itself—contrary to the popular assumption.

These critics ask what development for the Third World would look like were it not governed by the economic and political agendas of elites in the North. What would contemporary Southern sciences and technologies look like if these cultures were not forced to seek balance between, on the one hand, the traditional Northern romanticized view of Northern sciences and technologies and, on the other hand, the kinds of resistance to them that romanticize traditional knowledge systems of the South? Important foci in the critiques of development explore how to benefit from the very real resources both traditions offer while avoiding the costs of romanticizing either.

Multicultural science education.

Finally, in the last decade, a lively debate has arisen over if, to whom, and how non-Western systems of knowledge about the "natural world" (a Western concept, it turns out) should be taught in kindergarden through high school science classes (Cobern 2001; Stanley and Brickhouse 1994). The topic heated up in the late 1980s and early 1990s in New Zealand, where Maori educators had already convinced the federal government to include Maori achievements and perspectives in the new national standards for humanities and social science courses. Why not in the sciences, too, they asked. (SAME 1990ff) The multiculturalism issue began as the pedagogical concern to "start where students are" in multicultural classrooms in order better to recruit them into Western scientific training. It expanded into the curricular issue of widening the horizons for all students in their thinking about how cultures have successfully interacted with their environments and the scientific issue of learning to gather resources produced by the diversity of cultures to enable more effective interaction with nature's order. The similar U.S. debates started up in the early 1990s and now have an ongoing presence in science education journals, professional associations, and conferences. At issue are philosophic questions such as what is the fundamental nature of science, what does it mean to provide all students with an objective and balanced view of the scientific achievements of peoples of the world, and whether and how other cultures' sciences are valuable to them and to us.

These four intellectual and social movements offer additional resources for identifying the network of belief that persists in making it difficult to achieve a postpositivism that succeeds in abandoning problematic aspects of the unity-of-science thesis.

A World of Sciences vs. A World of Differences

I have been arguing that it is important to gain a richer appreciation of the empirical knowledge systems of other cultures, but that a robust philosophic engagement with these systems requires a more critical appreciation of the history, strengths, and limitations of our own knowledge systems. Yet, the possibilities of just such an appreciation can be advanced by using the new sociologies and histories of these other knowledge systems to identify problematic residual sources of support for the unity-of-science thesis.

Multicultural and postcolonial science studies provide a powerful critical perspective on each of those problematic residual "nodes" in the unity-thesis web of belief. The value of abandoning the forced choice between internalist or externalist histories and epistemologies of science is emphasized in accounts of the interdependency relations between the successes of European sciences and of European expansion: each made the other possible and gave distinctive cultural meaning to it. Thus, philosophies of science and their notions of objective observers, good method, the rationality of science, and the unity-of-science thesis itself are in part products of and contributors to such codependency relations. Moreover, a kind of logic of discovery is supplied by the multicultural and postcolonial accounts in their focus on how certain topics of inquiry and not others got to be interesting to European sciences (Brockway 1979; McClellan 1992). They engage in a standpoint project in this respect.

The symmetrical methodologies of the comparative ethnosciences projects undermine both exceptionalist and triumphalist assumptions about Western sciences. They expand the horizon of similar "deflationary" accounts in recent Northern science and technology studies. Finally, the usefulness of an interactionist instead of a representationalist conception of scientific output becomes especially clear when considering the effective interactions with nature's order achieved by other cultures' knowledge systems. Such an understanding goes a long way toward undermining the grounds for damaging forms of judgmental relativism, which thrive on issues about choosing the best scientific statement but wither when the issue is shifted to choosing the most effective or otherwise desirable practice.

Measures of effectiveness and desirability of practices must always be considered relative to goals; there is no possibility of an ideal of effectiveness or desirability in general of a scientific practice. Without the misleading conceptual device of a correspondence theory of scientific representation, it is hard for judgmental relativism to get a hold in our accounts of how sciences work.

I have been arguing that postpositivist philosophies of science, on the one hand, and multicultural and postcolonial science studies, on the other, are converging in important ways and that the former can benefit from insights about sciences in the North and in the South that emerge more easily from the latter studies. In particular, the implausibility of problematic residual elements of the unity-of-science thesis becomes more easily visible from the perspective of accounts of the ways histories and philosophies of science were generated in the North through Northern encounters with other cultures' knowledge traditions.

Such a critical understanding of our own traditonal knowledge system is valuable in itself. But it also opens up possibilities for just the kind of fruitful dialogues with other cultures that can continue to advance the growth of human knowledge. And, by no means least of all, it paves the way for richer explorations of how philosophies of science can both block and contribute to global social justice.[11]

Notes

1. "West" vs. "the rest" (or, originally, "the Orient"), and "first world vs. third world" are two of the conventional ways of contrasting the cultures of the world that figured in the "voyages of discovery" and then in the more recent last half century of development policies. Both contrasts are deeply problematic; the first retains the orientalist conceptual framework, and the second the categories of Cold War politics. The accuracy, and politics, of the contrast "developed vs. underdeveloped" has also been widely questioned (Sachs 1992). "North vs. South" became the preferred contrast at the 1992 United Nations' Rio de Janeiro Earth Summit. Of course, each of these contrasts also misassigns particular cultures to one side or the other. And the emphasis on such contrasts misleadingly homogenizes each side and overemphasizes differences between them. Yet recognition and discussion of macrodifferences between the "haves" and the "have-nots" should not disappear just when a critical focus on the history and practices of the "haves" appears. Moreover, particular discourses and institutions have gotten attached to one or another of such labeled contrasts while recognizing the problems with them. Thus, all four contrasts are still in play in postcolonial discussions, and I shall move between them as seems appropriate.

A question: What do I mean by democracy or "prodemocratic social relations" here and below? This question raises complex and vexed issues that currently are vigorously debated in political philosophy (see, for example, Benhabib 1996). Here I shall mean by it only a general ethic (John Dewey's proposal), namely, that those who bear the consequences of decisions should have a proportiate share in making them. Just which institutions, rules, and practices best advance such an ethic will contextually vary and will always be open to further debate.

2. As political theorist Val Plumwood pointed out to me in conversation.

3. Hacking (1996) distinguishes these two original senses of the unity of science.

4. Dupré (1993), for one, has challenged the reality and the ideal of a singular order of nature.

5. We need "strong objectivity," as I have argued elsewhere, e.g. Harding 1998.

6. Recent overviews of significant parts of it can be found in Harding 1998 and Hess 1995. See also Selin 1997.

7. Joseph Needham's (1954ff, 1969) studies of Chinese science are probably the most extensive early anti-Eurocentric account of another culture's inquiry systems, though he still held vestiges of an exceptionalist view of Northern science. The Selin (1997) *Encyclopedia* is a good place to start for a detailed state-of-the-art understanding of the richness and usefulness of other cultures' traditions and of the misconceptions of them created by trying to describe and evaluate them within the parameters used to think about Northern modern sciences and their philosophies. The ninety essays here have been written by distinguished scholars from around the world, including at least a few Western-trained philosophers of science. See also Joseph 1991; *Indigenous Knowledge and Development Monitor*; Kaptchuk 1983; Lach 1977; Maffie 2001; Verran 2001; Watson-Verran and Turnbull 1995; and Weatherford 1988. Haraway 1989 and Traweek 1988 provide well-known examples of comparative ethnographies that examine how particular sciences (primatology and high-energy physics, respectively) get embedded in other cultures in productive ways.

8. This concern with methodological symmetry can misleadingly look like the symmetry called for by the "strong programme" in the sociology of knowledge. David Bloor (1977) argued that scientific beliefs that were true or highly supported and those that turned out to be false or lacked such support should be explained in the same ways, rather than appealing only to nature's order and the power of logical thinking in the former case and only to social factors in the latter case. Yet, the comparison here is different in that the other cultures' inquiry practices and consequent belief systems do come with at least an important kind of high empirical support: they are the practice and belief systems that have enabled other cultures to survive—mathematical, botanical, agricultural, biological, pharmacological, medical, climatological, oceanographic, cartological, astronomical, calendric, technological, psychological, and sociological systems.

9. Important contributors to such studies include Blaut 1993; Brockway 1979; Crosby 1987; Goonatilake 1984, 1992; Headrick 1981; Kochhar 1992–93; Kumar 1991; McClellan 1992; Nandy 1990; Petitjean et al. 1992; Reingold and Rothenberg 1987; Shiva 1989.

10. I have recounted this argument in a number of places, for example in Harding 1998.
11. I'm indebted to Alison Wylie for helpful comments on an earlier draft and for direction to valuable sources.

References

Bass, Thomas. 1990. *Camping With the Prince and Other Tales of Science in Africa.* Boston: Houghton Mifflin.

Benhabib, Seyla, ed. 1966. *Democracy and Difference: Contesting the Boundaries of the Political.* Princeton, N.J.: Princeton University Press.

Blaut, J.M. 1993. *The Colonizer's Model of the World: Geographical Diffusionism and Eurocentric History.* New York: Guilford Press.

Bloor, David. 1977. *Knowledge and Social Imagery.* London: Routledge and Kegan Paul.

Braidotti, Rosi, et al. 1994. *Women, the Environment, and Sustainable Development.* Atlantic Highlands, N.J.: Zed.

Brockway, Lucille H. 1979. *Science and Colonial Expansion: The Role of the British Royal Botanical Gardens.* New York: Academic Press.

Cartwright, Nancy. 1995. "The Metaphysics of the Disunified World." *Philosophy of Science Association 1994,* Proceedings of the 1994 Biennial Meeting of the Philosophy of Science Association, Vol. 2. Ed. David Hull, Micky Forbes, and Richard M. Burian. East Lansing, Mich.: Philosophy of Science Association.

Cat, Jordi, Nancy Cartwright, and Hasok Chang. 1996. "Otto Neurath: Politics and the Unity of Science." In *The Disunity of Science: Boundaries, Contexts, and Power.* Ed. Peter Galison and David J. Stump. Palo Alto: Stanford University Press.

Caws, Peter. 1967. "Scientific Method," *Encyclopedia of Philosophy.* New York: Macmillan.

Cobern, William W., ed. 2001. "Special Issue on the Science Education/Multiculturalism Debate." *Science Education* 85(1).

Crosby, Alfred. 1987. *Ecological Imperialism: The Biological Expansion of Europe.* Cambridge, U.K.: Cambridge University Press.

Dupre, John. 1993. *The Disorder of Things: Metaphysical Foundations for the Disunity of Science.* Cambridge, Mass.: Harvard University Press.

Feyerabend, P.K. 1975. *Against Method.* London: New Left.

Galison, Peter, and David J. Stump, eds. 1996. *The Disunity of Science.* Palo Alto: Stanford University Press.

Goonatilake, Susantha. 1984. *Aborted Discovery: Science and Creativity in the Third World.* London: Zed.

———. 1992. "The Voyages of Discovery and the Loss and Rediscovery of the 'Other's' Knowledge." *Impact of Science on Society* (167):241–64.

Gross, Paul R., and Norman Levitt. 1994. *Higher Superstition: The Academic Left and Its Quarrels with Science.* Baltimore: Johns Hopkins University Press.

Hacking, Ian. 1983. *Representing and Intervening.* Cambridge, U.K.: Cambridge University Press.

———. 1996. "The Disunities of the Sciences." In *The Disunity of Science.* Ed. Peter Galison and David J. Stump. Palo Alto: Stanford University Press.

Hanson, N.R. 1958. *Patterns of Discovery.* Cambridge,UK: Cambridge University Press.

Haraway, Donna. 1989. *Primate Visions: Gender, Race, and Nature in the World of Modern Science.* New York: Routledge.

———. 1991. *Simians, Cyborgs, and Women: The Reinvention of Nature.* New York: Routledge.

———. 1997. *Modest_Witness@Second_Milenium.FemaleMan©Meets_OncoMouse™: Feminism and Technoscience.* New York: Routledge.

Harding, Sandra. 1998. *Is Science Multicultural? Postcolonialisms, Feminisms, and Epistemologies.* Bloomington: Indiana University Press.

———, forthcoming. "Should Accountability be Required in the Context of Discovery? Extending Philosophy of Science from its Restriction to Epistemology."

———, ed. 2003. *The Standpoint Reader.* New York: Routledge.

Headrick, Daniel R., ed. 1981. *The Tools of Empire: Technology and European Imperialism in the Nineteenth Century.* New York: Oxford University Press.

Hess, David J. 1995. *Science and Technology in a Multicultural World: The Cultural Politics of Facts and Artifacts.* New York: Columbia University Press.

Indigenous Knowledge and Development Monitor. ⟨http://www.nuffic.ni/ciran/ikdm.html⟩.

Joseph, George Gheverghese. 1991. *The Crest of the Peacock: Non-European Roots of Mathematics.* New York: I.B. Tauris.

Kaptchuk, Ted J. 1983. *The Web That Has No Weaver: Understanding Chinese Medicine.* New York: Congdon and Weed.

Kochhar, R.K. 1992–1993. "Science in British India, pts. I and II." *Current Science* (India) 63:11, 689–94; 64: 55–62.

Kuhn, Thomas S. 1970 (1962). *The Structure of Scientific Revolutions,* 2nd ed. Chicago: University of Chicago Press.

Kumar, Deepak. 1991. *Science and Empire: Essays in Indian Context* (1700–1947). Delhi, India: Anamika Prakashan and National Institute of Science, Technology, and Development.

Lach, Donald F. 1977. *Asia in the Making of Europe.* Vol. 2. Chicago: University of Chicago Press.

Maffie, James, ed. 2001. *Social Epistemology* 15(4). Special Issue, "Truth from the Perspective of Comparative World Philosophy."

McClellan, James E. 1992. *Colonialism and Science: Saint Domingue in the Old Regime.* Baltimore: Johns Hopkins University Press.

Nandy, Ashis, ed. 1990. *Science, Hegemony and Violence.* Delhi: Oxford.

Needham, Joseph. 1954ff. *Science and Civilization in China.* 7 vols. Cambridge: Cambridge University Press.

———. 1969. *The Grand Titration: Science and Society in East and West.* Toronto: University of Toronto Press.

Petitjean, Patrick, et al., eds. 1992. *Science and Empires: Historical Studies about Scientific Development and European Expansion.* Dordrecht: Kluwer.

Quine, W.V. O. 1960. *Word and Object.* Cambridge, Mass: MIT Press.

Reingold, Nathan, and Marc Rothenberg, eds. 1987. *Scientific Colonialism: A Cross-Cultural Comparison.* Washington, D.C.: Smithsonian Institution Press.

Rouse, Joseph. 1987. *Knowledge and Power: Toward a Political Philosophy of Science.* Ithaca, N.Y.: Cornell University Press.

———. 1996. *Engaging Science.* Ithaca, N.Y.: Cornell University Press.

Sabra, I.A. 1976. "The Scientific Enterprise." In *The World of Islam.* Ed. B. Lewis. London: Thames and Hudson.

Sachs, Wolfgang, ed. 1992. *The Development Dictionary: A Guide to Knowledge as Power.* Atlantic Highlands, N.J.: Zed.

SAME (Science and Math Education) *papers* 1990, 1991, 1992, 1993. Waikato, New Zealand: Centre for Science and Math Education Research.

Sardar, Z., ed. 1988. *The Revenge of Athena: Science, Exploitation, and the Third World.* London: Mansell.

Schuster, John A., and Richard R. Yeo, eds. 1986. *The Politics and Rhetoric of Scientific Method: Historical Studies.* Dordrecht: Reidel.

Selin, Helaine. 1992. *Science Across Cultures: An Annotated Bibliography of Books on Non-Western Science, Technology, and Medicine.* New York: Garland Publishing.

——— ed. 1997. *Encyclopedia of the History of Science, Technology, and Medicine in Non-Western Cultures.* Dordrecht: Kluwer.

Shapin, Steven. 1994. *A Social History of Truth.* Chicago: University of Chicago Press.

———, and Simon Shaffer. 1985. *Leviathan and the Air Pump.* Princeton, N.J.: Princeton University Press.

Shiva, Vandana. 1989. *Staying Alive: Women, Ecology, and Development.* London: Zed.

Smith, Linda Tuhiwai. 1999. *Decolonizing Methodologies: Research and Indigenous Peoples.* London: University of Otago Press.

Snow, C.P. (1959) 1964. *The Two Cultures: And a Second Look.* Cambridge, U.K.: Cambridge University Press.

Stanley, W.B., and Nancy W. Brickhouse. 1994. "Multiculturalism, Universalism, and Science Education." *Science Education* 78: 387–98.

Traweek, Sharon. 1988. *Beam Times and Lifetimes.* Cambridge, Mass.: MIT Press.

Turnbull, David. 1993. *Maps are Territories: Science Is an Atlas.* Chicago: University of Chicago Press.

Verran, Helen. 2001. *Science and an African Logic.* Chicago: University of Chicago Press.

Watson-Verran, Helen, and David Turnbull. 1995. "Science and Other Indigenous Knowledge Systems." In *Handbook of Science and Technology Studies.* Ed. S. Jasanoff, G. Markle, T. Pinch, and J. Petersen. Thousand Oaks, Calif: Sage. 115–39.

Weatherford, Jack McIver. 1988. *Indian Givers: What the Native Americans Gave to the World.* New York: Crown.

Williams, Michael. 1991. *Unnatural Doubt: Epistemological Realism and the Basis of Skepticism.* Oxford: Basil Blackwell.

Wylie, Alison. 1999. "Rethinking Unity as a 'Working Hypothesis' for Philosophy of Science: How Archaeologists Expoit the Disunities of Science." *Perspectives on Science: Historical, Philosophical, Social,* 7(3): 293–317.

To Walk in Balance: An Encounter between Contemporary Western Science and Conquest-era Nahua Philosophy

James Maffie

Does the predictive and manipulative success of contemporary Western-style science challenge the epistemological credibility of the philosophico-cultural perspective of the indigenous, Nahuatl-speaking peoples of central Mexico? Ideologues of Western-style science have long argued that such success demonstrates the epistemological inferiority, if not complete bankruptcy, of non-Western philosophies and modes of inquiry. In what follows, I take issue with this argument. I contend the predictive and manipulative success of Western-style science does not epistemologically challenge Conquest-era Nahua philosophy. Why not? Because contemporary Western scientific and Conquest-era Nahua inquiries embrace two alternative, epistemologically incommensurable epistemologies. In brief, they ask different questions, try to solve different problems, and pursue different ultimate values and goals. Consequently, one cannot evaluate Nahua inquiry by scientific norms, values, and goals without begging the question in favor of the epistemological legitimacy of those norms, values, and goals; and therefore, one cannot argue without begging the question that the predictive and manipulative success of contemporary Western-style science challenges, no less undermines, the epistemological credibility of Conquest-era Nahua philosophy and inquiry.

Section I of what follows briefly presents some background regarding Conquest-era Nahua. Sections II through IV discuss Nahua metaphysics, the problematic defining Nahua philosophy, and Nahua epistemology (respectively).[1] Section V reviews the epistemology of Western-style science. Finally, section VI compares Western-style science and Nahua philosophy and inquiry.

I. Who Are the Nahua?

The Nahuatl-speaking peoples of the High Central Plateau of Mexico originated in what is now northern Mexico and southwestern United States, migrating south in successive waves to the central Mexican highlands during the thirteenth and fourteenth centuries. Nahuatl is a member of the Uto-Aztecan linguistic family along with Ute, Hopi, and Comanche. Conquest-era Nahuatl-speakers included, among others, the Mexica (dubbed "Aztecs" by Europeans), Acolhuans, Texcocans, Tlacopans, Chalcans, Tepanecs, and Tlaxcaltecs. Due to their common language and culture scholars commonly refer to them as the Nahua, and to their culture as Nahua culture. Nahua culture flourished in the fifteenth- and sixteenth-centuries prior to 1521 (C.E.), the fall of the Mexica capital, Tenochtitlan, and official date of the Conquest. Nahuatl remains the most widely spoken indigenous language in Mexico today.[2]

II. Nahua Metaphysics

At the center of Nahua metaphysics is the view that there exists a single, vital, dynamic, vivifying, eternally self-generating-and-self-regenerating sacred energy, power, or force. The Nahua called this sacred energy *teotl*. Elizabeth Boone writes, "The real meaning of [*teotl*] is spirit—a concentration of power as a sacred . . . force."[3] Jorge Klor de Alva writes, "*Teotl* . . . implies something more than the idea of the divine manifested in the form of a god or gods . . . it signifies the sacred in more general terms."[4] *Teotl* transcends such (modernist) dichotomies as god vs. non-god, personal vs. impersonal, animate vs. inanimate, and alive vs. dead, and is therefore not properly understood in such terms.

Teotl's self-generation-and-regeneration is identical with its generation-and-regeneration of the universe. *Teotl* created, as well as continually recreates, permeates, and shapes the universe. That which humans commonly regard as the universe—for example, sun, earth, humans, trees, and animals—is generated by *teotl*, from *teotl*, as one aspect, facet, or moment of its eternal process of self-generation-and-regeneration. In short, *teotl* is more than the unified totality of things; *teotl* is identical to everything and everything is identical to *teotl*. Since the universe and its contents are identical with *teotl*, they also transcend such dichotomies as personal vs. impersonal, animate vs. inanimate, etc. As the single, all-encompassing life force of the universe, *teotl* also vivifies the universe and its contents. Lastly, *teotl* is both immanent and transcendent. It is immanent in that it penetrates deeply into every detail of the universe and exists within the myriad of created things; it is transcendent in that it is not exhausted by any single, existing thing.

Process, movement, becoming, and transmutation are essential attributes of *teotl* and hence the universe and its contents. *Teotl* is accordingly better understood as ever-flowing and ever-changing energy-in-motion rather than as a static entity, being, or thing. Since identical with *teotl*, the universe and its contents are thus also properly understood as ever-flowing and ever-changing energy-in-motion.

One of the primary ways by which *teotl*'s process, movement, and becoming present themselves is the ceaseless, cyclical oscillation of complementary paired opposites. Although essentially dynamic and devoid of any permanent order, the created universe is nevertheless characterized by an immanent equilibrium and rhythm: one provided and constituted by *teotl*. *Teotl* presents itself in multiple aspects, preeminent among which is duality. This duality takes the form of the endless opposition of mutually arising, interdependent, and complementary polarities that divide, alternately dominate, and explain the diversity, movement, and momentary structure of the universe. These include being and not-being, order and disorder, life and death, light and darkness, male and female, hot and cold, and active and passive.[5] Life and death, for example, are mutually arising, interdependent, complementary sides of the same process. Life arises from death, death, from life. The artists of Oaxaca presented this duality artistically by fashioning a double-faced mask, one-half fleshed (alive), one-half skeletal (dead).

Since *teotl* is essentially processual, it is properly understood neither as being nor not-being but as becoming. Similarly, *teotl* is properly understood as neither order nor disorder but as unorder. Being and not-being, like order and disorder, are two dialectically interrelated polarities and facets of *teotl*, and, as such, not strictly speaking predicable of *teotl* itself.[6] Indeed, this point applies generally to all the abovementioned dualities: life vs. death, male vs. female, and so on.

Teotl's untiring process of generating-and-regenerating the universe is also one of untiring self-transmutation. The universe is *teotl*'s self-transmutation—not its creation *ex nihilo*. The Nahua conceived this process in two closely interrelated ways. First, they conceived it artistically. *Teotl* is a sacred artist who endlessly fashions and refashions itself into and as the universe. A contemporary Nahua song-poem from the state of Veracruz, Mexico, reads:

I sing to life, to man
and to nature, the mother earth;

because life is flower and it is song,
it is in the end: flower and song.
(Quoted in and translated by Sandstrom [1991, 229])

The universe, in other words, is *teotl's in xochitl, in cuicatl* or "flower and song." The Nahua use the expression "*in xochitl, in cuicatl*" to refer specifically to the composing and performing of song-poems (which include flutes, drums, dancing, incense, and costume) and to refer generally to creative, artistic, and metaphorical activity (such as singing poetry and painting-writing). As *teotl's* "flower and song," the universe and its contents are *teotl's* grand, ongoing work of artistic-cum-metaphorical *self-presentation*, work of performance art, or "metaphor in motion" (as Markman and Markman (1989, xx) put it).

The Nahua simultaneously conceived *teotl's* process of self-transmutation in shamanic terms. The universe is *teotl's nahual*, that is, "disguise" or "mask." The Nahuatl word *nahual* derives from *nahualli*, which signifies a form-changing shaman. The continuous becoming of the universe and its myriad contents is *teotl's* shamanic self-masking and self-disguising.[7]

Teotl artistically-cum-shamanically transmutates and masks itself in a variety of ways: (1) the apparent thingness of existents, that is, the appearance of static entities such as humans, mountains, or animals. This is illusory, since one and all are merely aspects of *teotl's* sacred motion; (2) the apparent multiplicity of existents, that is, the appearance of distinct, independently existing entities such as individual humans, plants, or mountains. This is illusory since there is only one thing: *teotl*; and (3) the apparent distinctness, independence, and irreconcilable oppositionality of order and disorder, life and death, male and female. This is illusory since one and all are complementary facets of *teotl*.[8]

As an epistemological consequence of *teotl's* artistic-cum-shamanic self-masking, when humans ordinarily gaze upon the world, they misperceive *teotl* as an individual human, as male, and so on. In light of this, pre-Conquest Nahua *tlamatinime* ("knowers of things," sages, philosophers; *tlamatini* [singular]) routinely characterized earthly existence as consisting of pictures, images, and symbols painted-written by *teotl* on its sacred *amoxtli* (Mesoamerican papyrus-like paper).[9] For example, the *tlamatini* Aquiauhtzin characterizes the earth as "the house of paintings."[10] His contemporary, Xayacamach, writes, "Your home is here, in the midst of the paintings."[11] Like the images on *amoxtli* painted-written by human artists, the images on *teotl's* sacred *amoxtli* are fragile and evanescent. Nezahualcoyotl writes, "We live only in Your painting here, on the earth . . . we live only in Your book of paintings, here on the earth."[12] Finally, Tochihuitzin

Coyolchiuhqui writes, "We only rise from sleep, we come only to dream, it is *ahnelli* [untrue, unrooted, undisclosing], it is *ahnelli* [untrue, unrooted, undisclosing], that we come on earth to live."[13]

Nahua *tlamatinime* thus conceived the dreamlike illusoriness of earthly existence in epistemological, rather than ontological, terms. Illusion was not an ontological category as it was, say, for Plato. In Book VI of the *Republic*, Plato employed the notion of illusion: to characterize an inferior or lower grade of being, reality, or existence (namely, a semi-real realm of becoming); to distinguish this inferior grade of reality from a superior, higher one (namely, the Forms); and to deny that earthly existence is fully real. In contrast, Nahua *tlamatinime* employed the concept of illusion to make the epistemological claim that the natural condition of humans is one of unknowing—not the metaphysical claim that *teotl*'s mask and all earthly existents are ontologically distinct from *teotl*, ontologically inferior to *teotl*, and so not fully real. Humans normally misperceive and misconceive *teotl*; that is, they normally perceive and conceive *teotl*'s mask. Indeed, human unknowing is one and the same as *teotl*'s shamanic self-masking. The mask of unknowing which beguiles us as human beings is a function of our human point of view. It is not a metaphysical dualism inherent in the make-up of things.

Nahua metaphysical monism entails the metaphysical impossibility of humans seeing anything other than *teotl*. After all, *teotl* is the only thing which exists to be seen. This notwithstanding, humans normally misperceive and misunderstand what they see. How can this be? Humans normally perceive and conceive *teotl*, under a description, and hence do so in a manner that is *ahnelli*—that is, untrue, unrooted, inauthentic, concealing, and nondisclosing. For example, humans perceive and conceive *teotl* under a description and hence unknowingly as an individual human, as maleness, as death, and so on. When they perceive *teotl* under a description they perceive *teotl*'s *nahual* or self-disguise.

Nahua epistemology accordingly claims that the only way for humans to know *teotl* is to experience *teotl* without description. Humans know *teotl* by means of a mystical-style union between their hearts and *teotl* that enables them to experience *teotl* directly and hence without mediation by language, concepts, or categories. In so doing, humans know *teotl* through *teotl*. Although *teotl* is metaphysically immanent within human hearts, it is nevertheless epistemologically transcendent. Humans are not guaranteed knowledge of the sacred.

A deep metaphysical difference thus divides the underlying epistemological problematics of Nahua epistemology, on the one hand, and Western

Cartesian-style epistemology and Western-style science, on the other. The latter problematic typically understands the subject and object dualistically and their relationship in terms of a "veil of perception." The subject's access to the object is indirect, being mediated by appearances or representations of the object. The Nahua's epistemological problematic understands the subject and object monistically (since both are identical with *teotl*) and their relationship in terms of a mask—and masks in Mesoamerican thinking possess very different properties than veils!

In their study of Mesoamerican shamanism (in which sixteenth-century Nahua epistemology was deeply rooted and to which it remained closely related), Markman and Markman (1989, xx) write, masks "simultaneously conceal and reveal the innermost spiritual force of life itself." The half-life, half-death mask above both conceals and reveals a face that is simultaneously neither-alive-nor-dead-yet-both-alive-and-dead, and in so doing, simultaneously conceals and reveals the complementary duality of life and death. The mask does not represent, symbolize, or point to something deeper, something hiding behind itself, since the simultaneously neither-alive-nor-dead-yet-both-alive-and-dead face rests right upon the surface. Our access to the simultaneously neither-alive-nor-dead-yet-both-alive-and-dead face is not mediated by a veil or representation. Instead, it is fully present in seeing yet hidden under a description by our unknowing. After years of ritual preparation, Nahua *tlamatinime* were able to see the life-death mask "unmasked" as it were, and in so doing, discern the complementary unity and interdependence of life and death.

III. The Problematic Defining Nahua Philosophy

The Nahua regarded life on earth for human beings as one filled with pain, sorrow, and suffering. The earth's surface was itself an extremely treacherous place. Its name, *tlalticpac*, literally means "on the point or summit of the earth," suggesting a narrow, jagged place surrounded by constant dangers.[14] The Nahuatl proverb, "*Tlaalahui, tlapetzcahui in tlalticpac*" ("It is slippery, it is slick on the earth"), was said of a person who had lived a morally upright life but then lost her balance and fell into moral wrongdoing, as if slipping in slick mud.[15] Humans lose their balance easily on *tlalticpac* and thus repeatedly suffer misfortune and ill-being.

Nahua *tlamatinime* conceived the raison d'etre of philosophical speculation in terms of this conception of the human situation. They turned to philosophy for practicable answers to what they regarded as *the* defining question of human existence: "How can humans maintain their balance upon the slippery earth?" Together, this situation and question constitute the *problematic* which functions as the defining framework for Nahua inquiry (be it epistemological, moral, aesthetic, or prudential). At bottom, morally, epistemologically, prudentially, and aesthetically appropriate human activity are defined in terms of the goal of humans' maintaining their balance upon the slippery earth, and all human activities are to be directed towards realizing it. To the question "How can humans maintain their balance upon the slippery earth?" Nahua *tlamatinime* answered, "Humans must conduct every aspect of their lives wisely." To the question "What is the best path for humans to follow on the narrow, jagged surface of the earth?" they answered, "The balanced, middle path since it avoids excess and imbalance, hence misstepping and slipping, hence misfortune and ill-being."

Nahua sages conceived *tlamatiliztli* (wisdom, knowledge) in creative, performative, and practical terms rather than in propositional or theoretical terms. *Tlamatiliztli* consists of non-propositional "know-how"—not propositional "knowledge that." It involves knowing how to conduct oneself so as to make one's way safely upon the slippery surface of earth. How do humans become wise? They must become *neltiliztli*, that is, true, well-rooted, authentic, and non-referentially disclosing. Their intellectual, emotional, imaginative, and physical dispositions and behavior—in short, their entire lives—must become deeply and firmly rooted in, and hence disclosing of, the sacred (*teotl*).[16]

Tlamatiliztli consists of four, ultimately indistinguishable, aspects. First, it consists of the practical ability to conduct one's affairs in such a way as to attain some measure of equilibrium and purity—and hence some measure of well-being—in one's personal, domestic, social, and natural environment. Secondly, it consists of the practical ability to conduct one's life in such a way as to creatively participate in, reinforce, adapt, and extend into the future the way of life inherited from one's predecessors. Thirdly, *tlamatiliztli* consists of the practical ability to conduct one's life in such a way as to participate in the regeneration-cum-renewal of the universe. Finally, it consists of the practical "know-how" involved in performing ritual activities which: genuinely present *teotl*, authentically embody *teotl*, preserve existing balance and purity, create new balance and purity, and participate alongside *teotl* in the regeneration of the universe.

The Nahua universe was a "participatory universe" characterized by a "relationship of compelling mutuality" or "interdependence" between humans and universe.[17] They regarded this as an obvious consequence of the interrelatedness and oneness of all things. Not only does the universe causally affect humans, humans also causally affect the universe. Human actions promote either cosmic harmony, balance, and purity, on the one hand, or cosmic disharmony, imbalance, and impurity, on the other.

Tlamatiliztli also involves treating the universe in a morally responsible manner. Because humans owe their existence to the sacred (*teotl*), they are born indebted to the sacred and bear a moral-cum-religious obligation to participate in the renewal of the universe. They repay their debt by performing ritual activities such as "flower and song," autosacrifice (self-inflicted bloodletting), and sacrificing of plants, animals, and humans.[18] *Tlamatiliztli* entails not only knowing *how* to perform these ritual activities but also *where* and *when* to perform them: hence the importance of knowing the sacred-ritual calendar and landscape.[19]

The Nahua characterized persons, things, activities, and utterances equally and without equivocation in terms of *neltiliztli* and understood *neltiliztli* in terms of well-rootedness in *teotl*. That which is well-rooted in *teotl* is genuine, true, authentic, and well-balanced as well as non-referentially disclosing and unconcealing of *teotl*.[20] Created things exist along a continuum ranging from those that are well-rooted in *teotl* (i.e. *nelli*), and, hence, authentically present, and embody *teotl* as well as disclose and unconceal *teotl*, at one end, to those things that are poorly rooted in *teotl* (i.e. *ahnelli*) and hence neither authentically embody and present *teotl* nor disclose and unconceal *teotl*, at the other end. The former, which include fine jade and well-crafted song-poems ("flower and song"), enjoy sacred presence. The life-death mask above, for example, embodies and discloses *teotl* more authentically than would a mask presenting life and death as mutually exclusive contradictories.

The wise person enjoys sacred presence. She is stable, well-rooted, solid, authentic, genuine, and true. She has mastered the art of living well. Like a skilled mountaineer, she is able to maintain her balance and avoid slipping while walking upon the narrow, jagged summit of the earth.[21] Over the years she has fashioned her *in ixtli in yollotl* ("face and heart," or character) into a pure and precious work of art that embodies balance-and-purity.[22] She also possesses "a wise face and a strong, humanized heart," that is, one characterized by sound judgment and apt sentiment. Her emotional, imaginative, perceptual, cognitive, and physical activities embody and promote balance-and-purity as well as avert imbalance-and-impurity. The Nahua likened the sage to well-formed, unblemished jade and quetzal plumes—that is, to earthly things that enjoy sacred presence and authentically present, truly embody, and genuinely disclose *teotl*'s own balance-and-purity.

Well-rootedness, authenticity, acting truly—in short, walking in balance—possess ineliminable moral and aesthetic dimensions. Genuine, well-rooted humans are necessarily morally upright, straight, pure, and virtuous. They are careful, temperate, adept, respectful, responsible, composed, steadfast, and trustworthy. What's more, their "face and heart" (*in ixtli in yollotl*) is beautiful, like fine jade and quetzal plumes. In contrast, unrooted, false, and not-genuine humans are morally vicious, impure, and crooked. They are careless, mendacious, duplicitous, disrespectful, irresponsible, untrustworthy, and slippery. Their lives are miscreations of twisted imbalance, insanity, and mishapedness.

In sum, Nahua inquiry aims at the practical-cum-prudential-cum-moral-cum-aesthetic goal of walking in balance upon the earth, and wisdom consists of the practical ability to do so successfully. Keeping one's balance requires that one be well-rooted, authentic, true, pure, morally righteous, and beautiful. But it also requires that one know how to become knowledgeable.

IV. Nahua Epistemology

The above problematic also defines the raison d'etre of inquiry from the epistemological point of view. The aim of cognition from the epistemological point of view is walking in balance upon the slippery earth, and epistemologically appropriate inquiry is that which promotes this aim. Nahua epistemology does not pursue goals such as truth for its own sake, accurate representation, empirical adequacy, or manipulation and control; nor is it motivated by questions such as "What is the (semantic) truth about nature?" or "How can we master and bend the course of nature to our will?" As we've seen, *tlamatiliztli* is performative, not discursive; creative and participatory, not passive or theoretical; concrete, not abstract; a "knowing how," not a "knowing that."

Nahua *tlamatinime* conceived *tlamatiliztli* in terms of *neltiliztli*. Humans cognize wisely (knowingly) if and only if they cognize well-rootedly, and they cognize well rootedly if and only if their cognizing is well rooted in *teotl*. The Nahua understood well-rootedness in terms of burgeoning. Burgeoning and rootedness are both vegetal notions deriving from the organic world of agricultural existence. The flowers and fruit of corn burgeon from their seeds, soil, and roots, and in so doing, embody, present, and disclose the latter's qualities. Analogously, wise cognizing is organically rooted in (*nelli*) and generated by *teotl*. It is the flower of an organic-like process consisting of *teotl*'s burgeoning, unfolding, and blossoming within a person's heart. As the generative expression of *teotl*, *tlamatiliztli* is one of the ways *teotl* genuinely and authentically presents, discloses, and unconceals itself. Foolish (unknowing) cognizing, in contrast, is unrooted (*ahnelli*) in *teotl*. *Teotl* fails to burgeon, unfold, and unconceal itself within such cognizing.[23]

Humans come to know *teotl* using their hearts—not heads or brains. Situated between head and liver, the heart is uniquely qualified to achieve the proper balance between the head's reason and the liver's passion. As a consequence of ritual activities such as "flower and song," mortification, autosacrifice, and penitence, the movement of one's heart resonates in harmony with and eventually melds with the movement of *teotl*. When this occurs, one directly experiences the undifferentiated oneness and dynamism of *teotl*. *Teotl* burgeons up through one's heart, presenting and disclosing itself to and through one's heart.[24] Such hearts are said to be "*teotlized*" and to "hav[e] *teotl* [with]in them".[25] One knows *teotl* via direct acquaintance with and union between one's heart and *teotl*.

Because *teotl* is ineffable, the *teotlized* heart's experience and understanding of *teotl* are also ineffable and hence unmediated by language, concepts, and symbols. Language and concepts (along with their attendant divisions and distinctions) are aspects of *teotl*'s disguise or mask, and therefore contribute to humans' misperception and misunderstanding. In short, humans cannot know *teotl* by description.

In light of the foregoing, Nahua *tlamatinime* turned to "flower and song" (art, poetry, and music) to *present*—rather than *represent*—*teotl*.[26] The Nahua considered performing song-poems to be the highest form of human artistry and hence the best way for humans to present and disclose *teotl*. In performing song-poems humans most closely imitate and participate in *teotl*'s own sacred, creative artistry. Song-poems were thus considered the appropriate medium of sagely expression, and sages were perforce singer-poets.

Successfully performing knowledge-enhancing ritual activities requires (among other things) that cognizers be well rooted, well balanced, pure, authentic, and morally righteous, and that they possess such qualities as strength, self-control, moderation, modesty, humility, and respect. Humans must show humility and respect towards the sacred before the sacred will disclose itself to them. The Nahua regarded the preceding characteristics as both epistemological and prudential-cum-moral-cum-aesthetic. They not only help humans become knowledgeable and live wisely, they also help humans live morally, authentically, beautifully, and in balance. Humans cannot become knowledgeable of *teotl* without also becoming genuine, beautiful, and morally righteous (and vice versa). Indeed, the process of epistemological self-improvement is identical with that of pruden-tial-cum-moral-cum-aesthetic self-improvement.

The Nahua also understood the process of becoming knowledgeable in terms of *tlamacehualiztli*, or "the meriting of things". According to Burk-hart (1989, 142), *tlamacehualiztli* derives from the verb *macehua*, "to obtain or deserve what is desired."[27] Humans come to "merit"—that is, "deserve" or "be worthy of"—*tlamatiliztli* as a consequence of performing prescribed ritual activities. Humans and the sacred coexist in a moral interrelation-ship of reciprocity, and becoming knowledgeable involves a morally regu-lated exchange with the sacred. When humans behave in ritually prescribed ways, they may expect to be granted those things they have come to merit. *Tlamatiliztli* emerges as a consequence of moral-cum-epistemological in-teraction and coparticipation with the sacred.

Because *teotl* is not a lifeless object or thing, Nahua epistemology is bet-ter understood as an epistemology of a "Thou" than as an epistemology of an "It."[28] One approaches a "Thou" as another living individuality, as life confronting life, and one understands another "Thou" by acquaintance. One does not try to understand another "Thou" by abstract description or by subsuming it under general theories or laws. Knowing a "Thou" is morally regulated by reciprocity, respect, and humility, and consequently treating a "Thou" as a means to one's ends or as something to dominate and exploit results in cognitive folly.

Finally, although Nahua empirical inquiry did concern itself with suc-cessful prediction (most notably in matters astronomical), it did not do so with the aim of controlling or manipulating nature (as is commonly ar-gued to be the case with Western-style science).[29] Nahua empirical inquiry embraced the threefold aim of: (1) compliance or conformity, that is, shap-ing one's actions, thoughts, and behavior so as to be in harmony and bal-ance with the movements of *teotl*; (2) aligning oneself with preexisting pat-terns and forces of the cosmos so as to promote human and cosmic balance; and (3) actively coparticipating in the patterns and forces of the universe with the goal rewewing the cosmos. Seasonal and calendrical festi-vals (such as the New Fire Ceremony) did not consist of humans trying to bend the cosmos to their will but rather their trying to cooperate with the universe in order to renew the universe.

V. Western Science

Any comparative discussion of Western-style science is made difficult by the actual history of Western science as well as the variety of existing theories of Western science. As Larry Laudan has shown, how people have conceived and practiced science has changed dramatically over the last two millennia of that patch quilt we Westerners call "the history of science": from Thales, Aristotle, and Galen through Copernicus, Kepler, Galileo, and Newton to Darwin, Mach, Maxwell, Einstein, Bohr, and Bell.[30] For example, largely due to the influence of Plato and Aristotle, science was regarded from antiquity through the eighteenth century as a deductive enterprise in search of certainty. While indispensable within the context of discovery, experience and experiment were considered epistemologically dispensable within the context of justification, since incapable of yielding certainty. It has only been since the nineteenth century that scientists and philosophers of science have embraced induction as an epistemologically legitimate method and accepted that science yields probabilities rather than certainties.

Indeed, the current consensus among philosophers of science seems to be that science lacks an essence and that there is no such thing as "the" method of science that necessarily distinguishes science from other styles of inquiry.[31] Consonant with Dewey's and Quine's suggestions that science is "self-conscious common sense" (as Quine put it), Deloria, Jr., Harding, Goonatilake, Horton, Turnbull, and Watson-Verran (among others) argue that induction, deduction, appeal to past and present experience, and reasoning by practical "trial and error" are not the exclusive or distinctive province of Western scientists but are found in every mother's, farmer's, and artisan's daily repertoire. Human beings across cultures and history engage in orderly, systematic thought.[32]

Having noted this, however, we must add that there are nevertheless important contingent differences distinguishing Western-style science from non-Western cognitive practices. Not all humans practice science (in the Western sense), and not all humans are protoscientists (in the contemporary Western sense).[33] Even if the aforementioned methods are shared by all humans in their daily affairs, not all humans embrace those methods with equal degrees of time, energy, consistency, or ambition; practice them from within identical systems of background beliefs; utilize them to the exclusion of other cognitive methods (such as mystical intuition); try to solve the same kinds of problems with them; or apply them with the same purposes, values, or goals in mind.

Our present task is also made difficult by the variety of current theories of science. Exactly which science is it that allegedly challenges Nahua philosophy? Is it the rationalist's science that values demonstrative, apodictic truth[34]; the empiricist's that values prediction or anticipation of the future course of experience[35]; the pragmatist's that values manipulation and control[36]; or the scientific realist's that values maximizing (semantic) truth, correctly representing nature, or discovering the underlying causal mechanisms that correctly explain surface phenomena?[37]

The foregoing variety notwithstanding, Western scientists and theorists of science rarely, if ever, include among the aims of science explicitly moral, political, or aesthetic values such as justice, authenticity, walking in balance, beautiful character, or flourishing. Indeed, one of the ideological hallmarks of modern Western science is its self-avowed value neutrality and rejection of all moral, aesthetic, and political values as epistemologically subjective and deleterious. One pursues the truth and "lets the chips fall where they may" vis-à-vis people's feelings, well-being, or interests. Any attempt to incorporate moral or political values into scientific decision making is routinely condemned as corrupting of science, for example, as "Lysenkoism." When, in rare moments, scientists or theorists of science do acknowledge that scientists *qua* scientists do embrace values such as successful prediction, conservatism, fecundity, or simplicity, they quickly add that these values are "cognitive" or "pragmatic," and hence, pose no threat to science's value-neutrality.[38]

VI. Comparison and Conclusion

Does the predictive and manipulative success of contemporary Western-style science challenge the epistemological credibility of the philosophico-cultural perspective of the indigenous, Nahuatl-speaking peoples of central Mexico? Our answer will be more cogent if we remain neutral regarding the various theories of Western science surveyed above. Too often, comparisons of Western scientific and non-Western styles of inquiry undercut their own cogency by assuming a specific theory of science, which claims, for example, that science by definition pursues truth for its own sake, treats nature mechanically, or denies a role to the imagination.[39]

However, if Dewey, Goonatilake, Harding, Quine, Turnbull, and others are correct in claiming that the styles of reasoning employed by scientists are continuous with those employed by ordinary folk across history and culture, then where shall we look for those features distinguishing Western scientific and Nahua styles of inquiry? I propose we look to the ultimate values, purposes and goals motivating these styles of inquiry—what I shall call their *epistemological axiologies*. What does each posit as the ultimate goal(s) of inquiry? What does each posit as intrinsically valuable or worth pursuing in inquiry? Relative to what end(s) does each define evidence, warrant, or knowledge? Answering what questions does each treat as the raison d'etre of inquiry?

Depending upon whether one is a rationalist, scientific realist, empiricist, or pragmatist, the epistemological axiology of Western-style science defines the purpose and norms of inquiry in terms of such ultimate goals (intrinsic values) as: apodictic truth; correct description, representation or explanation; empirical adequacy; or successful control, respectively. Similarly, the respective raison d'etre of scientific inquiry is answering such problematic-defining questions as: "What is the truth about nature?"; "What are the underlying causal mechanisms of nature?"; "What regularities can we observe in nature?"; and "How can we successfully control the future course of experience?" (respectively). With the exception of pragmatist theories, these questions are standardly construed as theoretical, not practical; they concern what we should believe or accept, not how we should conduct our lives. With the same exception, Western-style scientific knowledge tends to be construed as propositional, discursive, and abstract; a "knowing that," not a "knowing how."

Nahua epistemological axiology conceives the purpose and norms of inquiry in terms of the ultimate goal of human beings maintaining their balance as they walk upon the slippery surface of the earth. The raison d'etre of inquiry is to provide practicable answers to its problematic-defining questions: "Where is the proper path for humans to follow?" and "How ought humans conduct themselves so as to maintain their balance (and hence the balance of the universe) while walking upon the earth?".[40] Nahua inquiry does not ask "What is the (semantically) true theory of nature?" or "What regularities can we observe in nature?" or "What are the underlying causal mechanisms of nature?" It aims neither at apodictic certainty, truth for truth's sake, representational accuracy, empirical adequacy, nor manipulative success. The Nahua's questions are practical, not theoretical. They concern how humans ought to live their lives. The answers to their questions are performative, not discursive. Nahua knowing is a "knowing how"—not a "knowing that." It is knowing how to cooperate with and participate alongside nature in the renewing of nature (and humankind as part of nature)—not how to manipulate nature. Nahua knowing is a morally regulated personal knowing by acquaintance—not an amoral, impersonal knowing by detachment, description, and abstraction.

In conclusion, Western-style science and Conquest-era Nahua inquiry ask different questions, try to solve different problems, and embrace different ultimate values and goals. In light of this, I submit they constitute two alternative, epistemologically incommensurable epistemologies.[41] Consequently, one cannot evaluate Nahua inquiry and philosophy by scientific norms and goals without begging the question in favor of the epistemological legitimacy of scientific norms and goals; and thus one cannot argue that the predictive and manipulative success of contemporary Western-style science challenges, no less undermines, the epistemological credibility of Nahua philosophy and inquiry without begging the question in favor of the epistemological legitimacy of scientific norms and goals. Bluntly put, Western science's ability to send humans to the moon does not demonstrate its epistemological superiority over Nahua inquiry or the philosophicocultural perspective of Conquest-era Nahua. Therefore, if we hope to bring these two styles of inquiry (and their respective epistemological axiologies) into the light of meaningful comparative evaluation, we must do so on nonepistemological grounds. In the end, I suggest doing so depends upon whether Nahua and Western European forms of life share any broader philosophical aims.

Notes

1. In what follows I aim to approximate Nahua philosophy in the era of the Conquest (1521). Given the limitations of sources and distance in time, I attribute the view broadly to the Nahua, although it may be more accurately attributed to the upper elite of priests, scholars, and educated nobility. Certainly, views differed among priests, warriors, merchants, farmers and artisans; men and women; dominant and subordinate city-states; and various regional and ethnic subgroups. I attribute the view broadly to the period of the Mesoamerican-European contact, realizing that cultures and philosophies are porous, living, works in progress. I employ Western philosophical categories such as axiology, metaphysics, and so on hermeneutically in order to introduce Nahua thought to readers schooled in Western philosophy—not for accuracy. These categories have no precise, uncontroversial equivalents in Nahua thought. The Nahua, for example, made no sharp distinction between what Western philosophers call ethics and epistemology. A more accurate understanding of Nahua thought thus requires that readers eventually reconceive these categories as a single unity. Research for this paper was made possible by a grant from the National Science Foundation distributed through the American Philosophical Association. For their input and support, I would like to thank James Boyd, Gordon Brotherston, David Carrasco, Robert Figueroa, Julie Greene, Sandra Harding, Grant Lee, Paul Roth, Alan Sandstrom, Ben-Ami Scharfstein, Helmut Wautischer, Alison Wylie, three anonymous referees, and the late David L. Hall. Special thanks go to Willard Gingerich.

2. Our sources for studying Conquest-era Nahua philosophy include precontact and early postcontact native pictorial manuscripts or "codices" (e.g., the *Codex Borbonicus* and *Codex Mendoza*), reports by the Conquerors, and ethnography-style chronicles composed by the first missionary friars entering Mexico after the Conquest. Friars Sahagun, Olmos, Motolinia, Duran, and Mendieta sought knowledge of Nahua culture and questioned the survivors of the Conquest about their culture. Friar Sahagun in particular assembled hundreds of folios containing enormous amounts of information that serve as the basis for his

Historia General de las Cosas de Nueva Espana and *Florentine Codex*. The *Cantares mexicanos* and *Romances de los Senores do Nueva Espana* consist of transcriptions of native song-poems compiled by natives under Spanish supervision during the last part of the sixteenth century. Recent ethnographies of contemporary Nahua also prove useful, e.g., Alan R. Sandstrom, *Corn Is Our Blood: Culture and Ethnic Identity in a Contemporary Aztec Indian Village* (Norman: University of Oklahoma Press, 1991) and Timothy J. Knab, *A War of Witches: A Journey in to the Underworld of the Contemporary Aztecs* (Boulder, Colo.: Westview, 1995), as do contemporary archaeological studies (e.g., Michael Smith, *The Aztecs* (Oxford: Blackwell, 1996). For discussion of Nahua language, history, thought, and culture, the social, cultural, and environmental context of Nahua philosophy, and our sources for studying the Conquest-era Nahua, see Robert M. Carmack, Janine Gasco, and Gary H. Gossen (eds.), *The Legacy of Mesoamerica: History and Culture of a Native American Civilization* (Upper Saddle River, N.J.: Prentice-Hall, 1996), James Maffie, " 'Like a Painting We Will Be Erased, Like a Flower, We Will Dry Up Here on Earth': Ultimate Reality and Meaning according to Nahua Thought in the Era of the Conquest," *Ultimate Reality and Meaning: Interdisciplinary Studies in the Philosophy of Understanding* 23 (December 2000):295–318; and Smith (1996).

3. Elizabeth Hill Boone, *The Aztec World* (Washington, D.C.: Smithsonian Books, 1994), 105. See also: Peter T. Markman and Roberta H. Markman, *Masks of the Spirit: Image and Metaphor in Mesoamerica* (Berkeley: University of California Press, 1989); H.B. Nicholson, "Religion in Pre-Hispanic Central Mexico." In *Handbook of Middle American Indians*, vol.10, ed. G. Ekholm and I. Bernal (Austin: University of Texas Press, 1971), 395–446; and Richard F. Townsend, *The Aztecs* (London: Thames and Hudson, 1972).

4. Jorge Klor de Alva, "Christianity and the Aztecs," *San Jose Studies* 5 (1979):7. This view appears to be shared by native North Americans. According to Sioux scholar Vine Deloria, Jr., "the most common feature of [tribal people's] awareness of the world [is] the feeling or belief that the universe is energized by a pervading power . . . that affects and influences them. . . . The presence of energy and power is the starting point of their analyses and understanding of the world. (Quoted in Barbara Deloria, Kristen Foehner, and Sam Scinta (eds.), *Spirit and Reason: The Vine Deloria, Jr., Reader* (Golden, Colo.: Fulcrum, 1999), 356).

   Native North Americans call this power *wakan orenda* or *manitou*. For further discussion, see Vine Deloria, Jr., *God Is Red: A Native View of Religion* (Golden, Colo.: Fulcrum, 1994); Mircea Eliade, *Patterns in Comparative Religion*, trans. Rosemary Sheed (Lincoln, Neb.: University of Nebraska Press 1966); D.M. Dooling and Paul Jordan-Smith (eds.), *I Become Part of It: Sacred Dimensions of Native American Life* (San Francisco: Harper-Collins, 1992); Leroy N. Meyer and Tony Ramirez, "Wakinyan Hotan: The Inscrutability of Lakota/Dakota Metaphysics." In *From Our Eyes: Learning from Indigenous People*, ed. S. O'Meara and D.A. West (Toronto: Garamound Press, 1966), 89–105.

5. See Alfonso Caso, *The Aztecs: People of the Sun*, trans. Lowell Dunham (Norman: University of Oklahoma Press, 1958); Louise M. Burkhart, *The Slippery Earth: Nahua-Christian Moral Dialogue in Sixteenth-Century Mexico* (Tucson:

University of Arizona Press, 1989); Miguel Leon-Portilla, *Aztec Thought and Culture: A Study of the Ancient Nahuatl Mind*, trans. Jack Emory Davis (Norman: University of Oklahoma Press, 1963); Alfredo Lopez Austin, *The Human Body and Ideology: Concepts of the Ancient Nahuas*, trans. Thelma Ortiz de Montellano and Bernard Ortiz de Montellano. 2 vols. (Salt Lake City: University of Utah Press, 1988); and Alfredo Lopez Austin, *Tamoanchan, Tlalocan: Places of Mist*, trans. Bernard Ortiz de Montellano and Thelma Ortiz de Montellano (Niwot, Colo.: University Press of Colorado, 1997).

6. For discussion of order, disorder, and unorder in Navajo metaphysics, see Gary Witherspoon, *Language and Art in the Navajo Universe* (Ann Arbor: University of Michigan Press, 1977); and in Taoist metaphysics, see R. Young and R. Ames, "Introduction to Ch'en Ku-ying," in *Lao Tzu: Text, Notes, & Comments.* trans. and adapted by R. Young and R. Ames (San Francisco: Chinese Materials Center, 1977).

7. See Peter T. Furst, "Shamanistic Survivals in Mesoamerican Religion," *Actas del XLI Congreso Internacional de Americanistas*, vol. III (Mexico: Instituto Nacional de Anthropologia e Historia, 1976), 149–157; Willard Gingerich, "*Chipahua-canemliztli*, 'The Purified Life,'" in the Discourses of Book VI, Florentine Codex." In *Smoke and Mist: Mesoamerican Studies in Memory of Thelma D. Sullivan*, Part II, eds. J. Kathryn Josserand and Karen Dakin (Oxford: British Archaeological Reports, 1988), 517–44; Markman and Markman (1988); Bernard R. Ortiz de Montellano, *Aztec Medicine, Health and Nutrition* (New Brunswick, N.J.: Rutgers University Press, 1990).

8. According to Sandstrom (1991, 138), for contemporary Nahua in Veracruz, Mexico,

> everybody and everything is an aspect of a grand, single, overriding unity. Separate beings and objects do not exist—that is an illusion peculiar to human beings. In daily life we divide up our environment into discrete units so that we can talk about it and manipulate it for our benefit. But it is an error to assume that the diversity we create in our lives is the way reality is actually structured . . . everything is connected at a deeper level, part of the same basic substratum of being. . . . The universe is a deified, seamless totality.

9. Although Nahua *tlamatinime* were not philosophers in the contemporary Euro-American sense of professional academics, they were "lovers of wisdom" who engaged in "thoughtful interaction with the world" (as Hester and McPherson put it) and thus philosophers in the traditional Greek sense of Socrates. For discussion of indigenous philosophers and philosophy, see Lee Hester, Jr., and Dennis McPherson, "Editorial: The Euro-American Philosophical Tradition and Its Ability to Examine Indigenous Philosophy," *Ayaangwaamizim: The International Journal of Indigenous Philosophy* 1 (1997):3–9, and James Maffie, "Editor's Introduction: Truth from the Perspective of Comparative World Philosophy," *Social Epistemology* 15 (December 2001):263–74.

10. *Cantares mexicanos*, fol.10 r., quoted in and translated by Miguel Leon-Portilla, *Fifteen Poets of the Aztec World* (Norman: University of Oklahoma Press, 1992), 282. Aquiauhtzin (ca.1430–1500) hailed from the hamlet of Ayapanco.

11. *Cantares mexicanos*, fol. 11v., quoted in and translated by Leon-Portilla (1992, 228). Xayacamach (second half of the fifteenth century) governed the town, Tizatlan, in Tlaxcala.

12. *Romances de los senores de Nueva Espana*, fol.35 r., quoted in and translated by Leon-Portilla (1992, 83). Nezahualcoyotl (1402–1472) was ruler of the city-state, Tezcoco.

13. *Cantares mexicanos*, fol. 10r., quoted in and translated by Leon-Portilla (1992, 221) (brackets by author following Willard Gingerich, "Heidegger and the Aztecs: The Poetics of Knowing in Pre-Hispanic Nahuatl Poetry," in *Recovering the Word: Essays on Native American Literature*, ed. B. Swann and A. Krupat (Berkeley: University of California Press, 1987), 85–112). Tochihuitzin Coyolchiuhqui (late fourteenth-century-early fifteenth-century) hailed from Tenochtitlan, capital city of the Mexica empire. For further discussion of *nelli* and *neltiliztli*, see James Maffie, "Why Care about Nezahualcoyotl?: Veritism and Nahua Philosophy," *Philosophy of the Social Sciences* 32 (March 2002): 73–93.

14. Translation by Michael Launey, quoted in Burkhart (1989, 58).

15. Fray Bernardino de Sahagun, *Florentine Codex: General History of the Things of New Spain*, eds. and trans. Arthur J.O. Anderson and Charles Dibble (Sante Fe, N.M.: School of American Research and University of Utah, 1953–1982:VI), 228, translation by Burkhart (1989).

16. For further discussion of *neltiliztli*, see Gingerich (1987) and Maffie (2002).

17. Johannes Wilbert, "Eschatology in a Participatory Universe: Destinies of the Soul among the Warao Indians of Venezuela," in *Death and the Afterlife in Pre-Columbian America*, ed. Elizabeth Benson (Washington, D.C.: Dumbarton Oaks, 1975), 163–89. See also Miguel Leon-Portilla, "Those Made Worthy by Sacrifice" in *South and Meso-American Spirituality*, ed. Gary Gossen in collaboration with Miguel Leon-Portilla (New York: Crossroad Publishing, 1993), 41–64; Lopez Austin (1997); Kay A. Read, "The Fleeting Moment: Cosmogony, Eschatology, and Ethics in Aztec Religion and Society," *Journal of Religious Ethics* 14 (1986):113–138; Sandstrom (1991); Richard F. Townsend, *State and Cosmos in the Art of Tenochtitlan*, Studies in Pre-Columbian Art and Architecture, no.20 (Washington, D.C.: Dumbarton Oaks, 1979); Deloria et al. (1999), 41–60; Deloria, Jr. (1994); Henri Frankfurt and H.A. Frankfurt, "Myth and Reality." In *The Intellectual Adventure of Ancient Man*, ed. Henri Frankfurt, H.A. Frankfurt, John Wilson, Thorkild Jacobsen, and William A. Irwin (Chicago: University of Chicago Press, 1946), 3–30.

18. For further discussion, see: Burkhart (1989); David Carrasco with Scott Sessions, *Daily Life of the Aztecs* (Westport, Conn.: Greenwood Press, 1998); Leon-Portilla (1963) and (1993); Read (1986); and Sandstrom (1991).

19. See Philip P. Arnold, *Eating Landscape: Aztec and European Occupation of Tlalocan* (Niwot, Colo.: University Press of Colorado, 1998); Carrasco and Sessions (1998); and Sandstrom (1991).

20. See Gingerich (1987) and (1988), and Maffie (2002).

21. This conception of wisdom is deeply rooted in indigenous Mesoamerican shamanism. See Burkhart (1989); Mircea Eliade, *Shamanism: Archaic Tech-*

niques of Ecstasy (Princeton, N.J.: Princeton University Press, 1964); Furst (1976); Gingerich (1988); and Myerhoff (1974).

22. Leon-Portilla (1963, 113–15). This discussion is indebted to Leon-Portilla (1963) and Lopez Austin (1988).

23. For further discussion, see Gordon Brotherston, *Image of the New World: The American Continent Portrayed in Native Texts* (London: Thames and Hudson, 1979), and James Maffie, "Flower and Song in the 'House of Paintings': A Philosophical Reconstruction of Nahua Epistemology at the Time of the Conquest" (unpublished manuscript).

24. As one song-poem puts it:

> From whence come the flowers that enrapture man?
> The songs that intoxicate, the lovely songs?
> Only from His home do they come, from the
> innermost part of heaven.
> (*Cantares Mexicanos*, fol. 34 r., quoted in and translated by Leon-Portilla (1963,77)).

25. Lopez Austin (1988, vol. I, 258ff., and vol. II, 245, 298); see also Leon-Portilla (1963).

26. Two song poems read:

> It is a true [nelli] thing, our song;
> it is a true [nelli] thing, our flowers,
> the well-measured song.
> (*Romances de los senores de Nueva Espana*, f.41, quoted in and translated by Gingerich (1987, 103) (brackets mine).)

> In flowers is the word
> of One God held secure.
> (*Cantares mexicanos*, f.11, quoted in and translated by Gingerich (1987, 103).)

27. See also Jorge Klor de Alva, "Aztec Spirituality and Nahuatized Christianity." In Gossen and Leon-Portilla (eds.) 1993, 173–97; Leon-Portilla (1993); and Read (1986). For a similar relationship among the indigenous peoples of North America, see Deloria et al. (1999), 40–61; and Jim Cheney and Anthony Weston, "Environmental Ethics as Environmental Etiquette: Toward an Ethics-based Epistemology," *Environmental Ethics* 21 (1999):115–34.

28. Although a "Thou," *teotl* is not an intentional agent in the modern Western sense. I borrow this terminology from H. Frankfurt and H.A. Frankfurt (1946); see also Deloria et al. (1999); and Deloria, Jr. (1994).

29. For discussion of Nahua astronomy, see Anthony Aveni, *Skywatchers of Ancient Mexico* (Austin: University of Texas Press, 1980); Johanna Broda, "Astronomy, Cosmovision, and Ideology in Pre-Hispanic Mesoamerica," in *Ethnoastronomy and Archeoastronomy in the American Tropics*, ed. Anthony F. Aveni and Gary Urton (New York: Annals of New York Academy of Science, vol. 385, 1982) 81–110; Carrasco and Sessions (1998); and Townsend (1992).

30. Larry Laudan, "The Demise of the Demarcation Problem." In *Beyond Positivism and Relativism* (Boulder, Colo.: Westview Press, 1996), 210–22. For additional discussion of the changing practices and conceptions of science, see E.A.

Burtt, *The Metaphysical Foundations of Modern Science* (Atlantic Heights, N.J.: Humanities Press, 1952); John Losee, *Philosophy of Science: A Historical Introduction*, 3rd ed. (Oxford: Oxford University Press, 1993); and Carolyn Merchant, *The Death of Nature* (San Francisco: Harper, 1980).

31. See: Laudan (1996); Dudley Shapere, "The Character of Scientific Change." In *Scientific Discovery, Logic, and Rationality*, ed. Thomas Nickles (Dordrecht: Reidel, 1978), 61–101; Dudley Shapere, "Method in the Philosophy of Science and Epistemology," in *The Process of Science*, ed. Nancy Nersessian (Dordrecht: Martinus Nijhoff, 1987), 1–39; Paul K. Feyerabend, *Against Method* (London: Verso, 1975); Paul Roth, *Meaning and Method in the Social Sciences* (Ithaca, N.Y.: Cornell University Press, 1987); and Richard Rorty, *Consequences of Pragmatism* (Minneapolis: University of Minnesota Press, 1982).

32. See: John Dewey, *Reconstruction in Philosophy*, 2nd ed. (Boston: Beacon Press, 1948); Eliade (1966); W. V. O. Quine, "Two Dogmas of Empiricism," in *From a Logical Point of View*, 2nd ed. (New York: Harper & Row, 1951), 20–46; Susantha Goonatilake, *Toward a Global Science* (Bloomington: Indiana University Press, 1998); Sandra Harding, *Is Science Multicultural?: Postcolonialisms, Feminisms, and Epistemologies* (Bloomington: Indiana University Press, 1998); Deloria et al., (1998); Robin Horton, "Tradition and Modernity Revisited," in *Rationality and Relativism*, ed. Martin Hollis and Steven Lukes (Cambridge, Mass.: MIT Press, 1982), 201–60; Roth (1987); David Turnbull, "Local Knowledge and Comparative Scientific Traditions," *Knowledge and Policy* fall/winter (1993–1994):29–54; and Helen Watson-Verran and David Turnbull, "Science and other Indigenous Knowledge Systems," in *Handbook of Science and Technology Studies*, ed. S. Jasanoff, G.E. Markle, J.C. Petersen, and T. Pinch (London: Sage, 1995), 115–39.

33. See note 32.

34. See Burtt's (1952) discussions of Copernicus, Galileo, and Kepler.

35. See Bas Van Fraassen, *The Scientific Image* (Oxford: Oxford University Press, 1980).

36. See: Rudolph Carnap, "Empiricism, Semantic, and Ontology," in *Meaning and Necessity* (Chicago: University of Chicago Press, 1947), 205–21; Larry Laudan, *Science and Values: The Aims of Science and their Role in Scientific Method* (Berkeley: University of California Press, 1984); Quine (1951); W.V.O. Quine, *Ways of Paradox and other Essays*, 2nd ed. (Cambridge, Mass.: Harvard University Press, 1976); and Roth (1987).

37. See Richard Boyd, "The Current Status of Scientific Realism," in *Scientific Realism*, ed. Jarrett Leplin (Berkeley: University of California Press, 1984), 41–82; Philip Kitcher, *The Advancement of Science* (Oxford: Oxford University Press, 1993); and Ronald Giere, *Explaining Science: A Cognitive Approach* (Chicago: University of Chicago Press, 1988).

38. Having said this, one must quickly add that the preceding image of science has long been contested by Marxist, feminist, and third-world philosophers who argue that such cognitive values are in the final analysis, if not themselves, moral, aesthetic, or social-political, then at least firmly rooted in moral, aesthetic, or social-political values (e.g. those of capitalism, patriarchy, or imperialism). See Sandra Harding, *The Science Question in Feminism* (Ithaca, N.Y.:

Cornell University Press, 1986); Harding (1998); Sandra Harding (ed.), *The "Racial" Economy of Science: Toward a Democratic Future* (Bloomington: Indiana University Press, 1993); Evelyn Fox Keller, *Reflections on Gender and Science* (New Haven: Yale University Press, 1984); Evelyn Fox Keller and Helen Longino (eds.), *Feminism and Science* (Oxford: Oxford University Press, 1996); Ashis Nandy (ed.), *Science, Hegemony and Violence: A Requiem for Modernity* (Delhi: Oxford University Press, 1990); Peter Railton, "Marx and the Objectivity of Science," in *PSA 1984*, ed. Peter Asquith and Philip Kitcher (East Lansing, Mich.: Philosophy of Science Association, 1985), 813–25; Ziauddin Sardar (ed.), *The Revenge of Athena: Science, Exploration and the Third World* (London: Mansell, 1988); Third World Network, *Modern Science in Crisis: A Third World Response* (Penang: Third World Network, 1988).

39. Two prominent examples are Vine Deloria, Jr., and Robin Horton who base their respective evaluations of scientific vs. nonscientific thought upon outmoded positivist conceptions of science. See Deloria et al. (1999) and Robin Horton, "African Traditional Thought and Western Science," in *Rationality*, ed. Bryan Wilson (New York: Harper & Row, 1970), 131–71.

40. For example, Vine Deloria, Jr. writes,

> The old indians were interested in finding the proper moral and ethical road upon which human beings should walk. All knowledge . . . was directed toward that goal. The real interest of the old indians was not to discover the abstract structure of physical reality but rather to find the proper road along which, for the duration of a person's life, individuals were supposed to walk
> (Deloria *et al.*, 1999, 41–46).

41. The two remain axiologically nonequivalent even if one adopts nonmainstream feminist, Marxist, or Third World views of science which contend that the goals of Western-style science are, in the final analysis, the domination, control and exploitation of natural or human resources. As we've seen, these are not the goals of Nahua inquiry. See note 38.

Seeds and their Sociocultural Nexus

Hugh Lacey

Recent developments in agrobiotechnology undergird far-reaching changes in agricultural practices that are based on the development and practical utilization of transgenics. These developments depend upon successfully engineering new kinds of seeds which are not the products of crossbreeding, but which have been engineered so that their genomes contain genetic materials obtained from unrelated organisms. Although their roles in agricultural practices and their effects vary with social, economic, and cultural context, a widely publicized argument has been mounted that these developments serve human interests universally: they are indispensable for producing sufficient quantities of nourishing food to feed the world's increasing population over coming decades. Thus, it is said, transgenic seeds should be considered objects of high value for all value outlooks. Against this, I will contend, what seeds (used in agriculture) and the plants that grow from them are is partly a function of the sociocultural nexus (SCN) of which they are constituents and that their value does not significantly transcend their specific nexus. I will argue that the following two questions cannot be separated: How are seeds (plants and crops) to be scientifically investigated? How is the knowledge obtained from such investigations, on application, to be evaluated? The answers, in turn, vary with the SCN.

What Seeds Are

Seeds used in agriculture are simultaneously many things, including:

(a) Biological entities: under appropriate conditions they will grow into mature plants from which, for example, grain will be harvested.
(b) Constituents of various ecological systems.

(c) Entities that have themselves been developed and produced in the course of human practices.

(d) Objects of human knowledge and empirical investigation. As biological entities, they are subject to genetic, physiological, biochemical, cellular, developmental analyses; as parts of ecological systems, to ecological analyses; and as products of human practices, to analyses of their roles and effects in the SCN in which they are planted and their products distributed, processed, consumed and put to other uses.

The specific ways in which seeds are each of the above kinds of entities, and the specific possibilities that are open to them, vary systematically with the SCN of farming. Seeds used in farming may be, and traditionally have usually been, biological entities that are reproduced simply as part of the crop harvested. As such, they are renewable regenerative resources that (conditional upon a measure of social stability and absence of catastrophes) may be integral parts of sustainable ecosystems that generate products that meet local needs while being compatible with local cultural values and social organization and that have been selected by numerous farmers over the course of centuries with methods informed by local knowledge.[1] Traditionally, such seeds have been considered the common patrimony of humankind, available to be shared as resources for replenishing and improving the seeds of fellow farmers. In contrast, seeds may be commodities: objects bought and sold on the market, "property" whose users may not be their owners, whose features and uses are integrally connected with the availability of other commodities (e.g., chemical inputs and machinery for cultivation and harvesting), and that sometimes can be patented and otherwise regulated in accord with Intellectual Property Rights (IPR). Under these conditions, they are developed by professional breeders and scientists and produced largely by capital-intensive corporations. Then, they cannot be understood simply (and sometimes not at all) as part of the grain harvested, as components of stable ecosystems, and certainly not as entities to be freely shared with fellow farmers.

Increasingly throughout recent history, seeds have been transformed from being predominantly regenerative resources to commodities.[2] The transformation was initiated with the introduction of "high-intensity models" into agriculture, models based on mechanization and the use of extensive chemical inputs (fertilizers, pesticides, herbicides, etc.), and then further developed by the use of monocultures, of hybrid seeds that do not reproduce themselves reliably and so must be bought regularly from the seed company, and most recently, by the rapidly expanding use of transgenic seeds and the protections of IPR that they have been granted. The commoditization of the seed, which depends on breaking the unity of seed (on the one hand) as source of a crop and (on the other hand) as reproducer of itself (Shiva 1997), is an integral part of the transformation of the social relations of farming in the direction of the growing dominance of agribusiness and large scale farming, with, in many Third World countries, export orientation. It serves corporate interests. Its proponents also maintain that it enables greater efficiency in agriculture, and, above all, that the farming methods associated with it enable much greater and cheaper production of the grains needed to feed the world's growing population. It serves, they maintain, not only corporate interests, but also interests pertaining to all value outlooks.[3]

The Value of Transgenics

These issues come to a head in current controversies about transgenics. Do the development of transgenic seeds and the implementation of transgenic-intensive (TI) agriculture—as well as previous developments of "high-intensity models"—in fact serve human interests universally? Or, are there some value outlooks whose interests require enhancing forms of agriculture in which seeds remain predominantly regenerative resources?

Pro arguments for the universal value of TI agriculture tend to draw upon premises like the following:

1. Technology, informed by scientific knowledge, provides the unique key to solving major world problems like hunger and malnutrition.
2. Seeds are essentially the way in which they are represented in molecular, genetic, physiological, and cellular biology—in biotechnology.
3. The knowledge that informs developments of transgenics is characteristic scientific knowledge.
4. That knowledge may be applied, in principle evenhandedly, to serve the interests and to improve the practices of groups holding a wide variety of value outlooks.
5. There are great benefits to be had from transgenics-intensive agriculture now, and they will be greatly expanded with future developments of transgenics, which promise, for example, crops with enhanced nourishing qualities that may readily be grown by poor Third World farmers.
6. The transgenic crops that are currently being planted, harvested, processed, and consumed, and those anticipated, occasion no foreseeable risks concerning human health and the environment that

cannot be adequately managed under responsibly designed regulations.

7. TI agriculture is necessary to ensure that the world's expected population in the coming decades can be adequately fed and nourished. There are no other ways that are informed by the soundly accepted results of scientific investigation that can be counted on to produce (or even to play a large-scale subordinate role in producing) the necessary food.

This argument is difficult to resist. I cannot address it comprehensively here, but will restrict my attention to questioning Premise 7. This will also involve some questioning of Premises 2 to 4. Much effort has been expended by the proponents of TI agriculture defending Premises 5 and 6 against well-known criticisms. While many important questions arise in these disputes, I think that issues surrounding Premise 7 are more fundamental.[4]

Premises 2 to 4 encapsulate widely held views about the nature of scientific inquiry. They deny that seeds as implicated in a SCN have much to do with how they are investigated as biological objects; biology is sharply separated from studies concerning the SCN. Whether particular seeds are commodities, renewable resources or gifts, sources of marketable products or foodstuffs for local consumption, grown for the sake of multiple products or a single one, seeds (and plants)—as objects of biological investigation—are effectively reducible to their genomes and to the biochemical expressions of their component genes. Their possibilities are encapsulated in terms of their generability from their underlying molecular structures (and the possibilities for their modification) and lawful biochemical processes. Seeds are essentially as they are investigated in molecular, genetic, physiological and cellular biology, that is, in biotechnology (Premise 2).[5]

Understanding seeds biologically in this way thus largely abstracts the realization of their possibilities from their relations with social arrangements, with human lives and experience, with the social and material conditions of the research, and with extensive and long-term ecological impact (and with any other beings that might be recognized in a culture's cosmovision)—thus, from any link with value. In turn, biological knowledge (so understood) is considered available to inform, more or less evenhandedly, agricultural practices regardless of the SCN in which they may be inserted (Premise 4). Whatever seeds may become in agroecosystems (sustainable or high-intensive) is determined by the possibilities that are encapsulated in their genomes and the possibilities of their transformation (whether by natural or farmer-directed selection, or by bioengineering) and realized by means of chemical interactions with substances encountered in their immediate environments. There can be no feedback into (basic) biological investigation from considerations pertaining to seeds as they are located in a SCN.

Agrobiotechnology (BT) utilizes mainstream and cutting-edge science (Premise 3). As such, its research is conducted under instances, agrobiotechnological strategies (BTS), of what I have called *materialist strategies* (MS).[6] Under MS, in the first place, theories are constrained to those with the lexical, mathematical, and modelling resources to be able to formulate posits of underlying order—structures and their components, processes, and interactions, and the laws that govern them. These theories identify the possibilities of phenomena (I call them *abstracted possibilities*) in terms of the generative power of the underlying order, in abstraction from any place they may have in human experience and practical activity, from any links with social value and with the human, social, and ecological possibilities that they might also admit. Although the abstracted possibilities of phenomena include possibilities that are in fact identical with possibilities for technological application, under MS they are investigated as abstracted possibilities not as social objects of any kind. In the second place, and reciprocally, under MS, data are selected so that their descriptive categories are generally quantitative, devoid of the categories of intentionality and value, applicable in virtue of measurement, instrumental, and experimental operations.

Are Transgenics Necessary to Feed the World?

What light does scientific investigation cast upon Premise 7? First of all, it is clear that research on transgenics and, in general, that conducted exclusively under MS casts no light on it. That kind of research can illuminate such things as the possibilities that may be open to genetic engineering and their potential impact on the character and quantity of production. At most, it can confirm that TI farming is a way to produce the food needed to feed the world—but potential other ways do not fall within its compass.[7]

Are there viable alternative forms of agriculture that can play an integral part in producing the food necessary to feed the world's growing population? Consider:

> It is not clear which are greater—the successes of modern high-intensity agriculture, or its shortcomings. . . .The successes [e.g., of the Green

Revolution] are immense. . . . But there has been a price to pay, and it includes contamination of groundwaters, release of greenhouse gases, loss
of crop diversity and eutrophication of rivers, streams, lakes and coastal
marine ecosystems. . . . It is unclear whether high-intensity agriculture
can be sustained, because of the loss of soil fertility, the erosion of soil,
the increased incidence of crop and livestock diseases, and the high energy and chemical inputs associated with it. The search is on for practices
that can provide sustainable yields, preferably comparable to those of
high-intensity agriculture but with fewer environmental costs, . . . that
incorporate accumulated knowledge of ecological processes and feed-
backs, disease dynamics, soil processes and microbial ecology.[8]

Tilman (1998)[9] describes some recent experimental studies that support
the promise of such "ecological" ("organic") alternatives, and Zhu et al.[10]
demonstrate that "a simple, ecological approach to disease control can be
used effectively at large spatial scale to attain environmentally sound disease control" without loss of productivity (compared to chemically intensive farming using monocultures). These studies complement Lewontin's
theoretical argument that there are methods, in continuity with traditional
farmer-selection methods, of "simple direct selection of high-yielding
plants in each generation and the propagation of seed from those selected
plants" that would enable "plant breeders [to], in fact, produce varieties of
corn that yield quite as much as modern hybrids" (Lewontin 1992).[11] Apparently, farming in which seeds are constituents of sustainable agroecosystems is not necessarily deficient in productivity.

   Shiva (1991) complements Tilman's and Lewontin's contentions in several ways. First, she[12] points to the productivity, potential for increased productivity, and agroecological soundness of many traditional agricultural
practices. Secondly, she questions the efficiency of the Green Revolution
(compared with potential developments of traditional methods) in view
both of the extensive and expensive chemical and other inputs needed to
produce the higher yields, and of (she alleges) exaggerated claims about productivity gains, since the actual gains made concern only a single crop and
have been achieved at the expense of reductions in other products of traditional farms. Thirdly, she adds an array of social consequences to the shortcomings of GR, listed by Tilman, including: displacement of traditional
small-scale farming, causing social dislocation (and consequent violence)
and hunger among the communities that sustained it; loss of the knowledge
that informs that kind of farming; and deepened dependence of Third World
conditions and possibilities on the interests of the global market.

   If there are alternatives, with characteristics as cited by Tilman or
Lewontin or Shiva, then the value of the use of "high-intensity models," in-

cluding TI agriculture, cannot be assured universally, especially among those groups who experience the sufferings induced in the light of the social shortcomings just listed. Even if TI agriculture may be the only form of high-intensity agriculture that is viable in the long term, there are few who expect it to produce significantly more high-yielding plants than those currently used in high-intensive farming, or to be more environmentally sustainable than well-designed ecological methods; so, whatever its other merits might be, they do not preclude that ecological methods may have higher value for some value outlooks. Shiva (1997) anticipates that implementations of TI agriculture will exacerbate the above shortcomings, and she emphasizes their (alleged) inability to provide solutions to the actual problems of small-scale farmers.[13] If so (Lacey 2001b) then, at least among those who bear the brunt of the shortcomings, transgenics will not be highly valued, especially if these people belong to movements aiming to develop alternate modes of farming that are highly productive, ecologically sustainable, and protective of biodiversity (Tilman) and are also compatible with social and cultural stability and diversity or with the values called "sustainability" below (Shiva 1997 and agroecologists, e.g., Altieri 1995). Furthermore, it has been maintained, the implementation of TI agriculture contributes to undermining such movements and their projects—by means of reliance on IPR claims, furthering the process of commoditization, and engagement in "biopiracy," that is, the free appropriation (sanctioned by law) of the seeds and knowledge of traditional cultures for commercial exploitation that, in turn, contributes to undermining the continued maintenance of seeds as regenerative resources.[14]

Agroecological Strategies

Although suggestive, none of this establishes decisively that there are alternatives that are informed by the soundly accepted results of scientific investigation. But they point to a limitation in the formulation of Premise 3: the knowledge that informs transgenics is indeed characteristic of knowledge gained under MS, but not of science in general. There is no available evidence that the possibilities of things can, in general, be reduced to their abstracted possibilities. The possibilities of seeds in sustainable agroecosystems (AE systems), for example, are not reducible to the possibilities that may be identified in investigations conducted under BTS (or other versions of MS). This does not preclude that research aiming to identify them may be conducted in systematic, empirical ways under other strategies that I call *agroecological strategies* (AES) (Lacey 1999, 2001a).

AES are particular instances of general ecological strategies that enable us to identify the possibilities that things (e.g., seeds) have in virtue of their place in AE systems.[15] Just as MS and other strategies do, AES require constraints on the kind of understanding to be sought—specification of the kinds of explanations to be developed and of possibilities to be identified—and selection of the features of empirical data that are to be sought and recorded (Lacey 1999). Under AES, research aims to confirm generalizations concerning the tendencies, capacities, and functioning of AE systems, their constituents, and relations and interactions among them. These include generalizations in which, for example, "mineral cycles, energy transformations, biological processes and socio-economic relationships" are considered in relationship to the whole system; generalizations concerned not with "maximizing production of a particular system, but rather with optimizing the agroecosystem as a whole" and so with "complex interactions among and between people, crops, soil and livestock."[16] Of particular salience are generalizations that help to identify the possibilities for sustainability of agroecosystems, where "sustainability" has been defined in terms of four interconnected characteristics: *Productive capacity*: "Maintenance of the productive capacity of the ecosystem"; *Ecological integrity*: "Preservation of the natural resource base and functional biodiversity;" *Social health*: "Social organization and reduction of poverty"; *Cultural identity*: "Empowerment of local communities, maintenance of tradition, and popular participation in the development process."[17]

Empirical data are selected and sought out, under AES, in virtue of their relevance for testing such generalizations and for enabling phenomena, relevant in light of sustainability, to be brought within the compass of investigation and application. Obtaining the data often requires subtle, regular, painstaking, accurate observation and monitoring of a multiplicity and heterogeneity of details in the AE systems. The skills for this are often developed principally by local farmers themselves, so that obtaining the data depends on the collaboration of farmers and the utilization of their experience and knowledge. Relevant data are often obtained from the study of farming systems in which traditional methods informed by traditional local knowledge are used. These systems are appropriately submitted to empirical scrutiny because AE studies have shown "that traditional farming systems are often based on deep ecological rationales and in many cases exhibit a number of desirable features of socioeconomic stability, biological resilience and productivity" (Altieri 1987, xiii; for examples, see Altieri 1995, ch. 6). They can, with adaptations suggested by research findings (e.g., those of Tilman and others), be enhanced with respect to all four of the characteristics of sustainability; and, especially with respect to cultural identity, they are often uniquely appropriate for the activities of poor, small-scale farmers. The methods used in these systems have been tested empirically in practice and have been particularly effective over the centuries in "selecting seed varieties for specific environments" (Altieri 1995, 116)—these are often the source of the seed varieties from which transgenics are engineered (Shiva 1997; Lacey 1999, ch. 8).[18]

Strategies and Their Links with Social Values

Research conducted under AES cuts across the strictures of MS. To many defenders of Premise 3, agroecology does not count as another way for it is not really informed by "science,"[19] where "scientific" research is identified with that conducted under MS. By definition, then, research conducted under AES is held to be "not scientific," for the biology is not separated from (though it is not determined by) the SCN, and empirical investigation aiming to further the embodiment of particular social values ("sustainability") is explicitly conducted. A priori, agroecology does not provide "another way" that could refute Premise 7. But this is to trivialize Premise 7, so much so, that any role it has in the legitimation of TI agriculture is lost. For the relevant question is surely: Is there empirical evidence that, apart from practices informed by knowledge gained under MS and especially its most advanced forms (transgenics), there are no alternative forms of agriculture—informed by systematic empirical inquiry—that can play a big part in meeting the world's food needs? It seems preferable to me to identify "science" not with research conducted under MS but with any systematic, empirical inquiry aiming to gain understanding of phenomena (Lacey 1999, ch. 5; 2000).

To be sure, the proponents of Premise 7 (and 3) are not alone in tending to identify "science" with research conducted under MS. Modern natural science has in fact been conducted almost exclusively under MS, and the spokespersons of its tradition have seldom recognized that, in many domains, this represents a choice. Why has modern science adopted (varieties of) MS almost exclusively? Why, in contemporary mainstream agricultural science, are AES largely ignored? I have discussed these questions in detail elsewhere (Lacey 1999, ch. 6; 2001a) and will only repeat my own answer here. MS are adopted almost exclusively because (in addition to being fruitful) adopting them has mutually reinforcing relations (described in Lacey, 1999) with *specifically modern ways of valuing the control of natural objects* (MVC). These values concern the scope of control, its

centrality in daily life, its relative unsubordination to other moral and social values—so that, for example, the kind of ecological and social disrup-
tion caused by high-intensity farming referred to above can be seen simply
as the price of progress. MVC also include the deep sense that control is the
characteristic human stance towards natural objects—so that the expansion of technologies (informed by knowledge gained under MS) into more
and more spheres of life and into becoming the means for solving more
and more problems is highly valued.

Commitment to MVC, further reinforced by contemporary global-
market institutions and policies that highly embody MVC, I suggest,
largely explains confidence in the possibilities of transgenics to solve major
problems of the poor, and thus prioritizes agricultural research conducted
principally under BTS. It also explains the ease with which the proponents
tend to paint all their critics with the anti-science ("luddite") brush (Premise 1).[20] But there is no scientific imperative to adopt BTS (or MS); the
decisive factor is commitment to specific (albeit hegemonic) social values:
MVC and/or those of the global-market. Premise 7 currently is grounded
in commitment to MVC; it is not the outcome of sound empirical sociohis-
torical inquiry. It may yet come to be vindicated empirically; but this cannot happen unless the limits of the productive capacity of agroecology are
tested severely, and so it cannot be vindicated by research conducted virtually exclusively under BTS (or MS).

Adopting either BTS or AES can be grounded in links with social values; the former with MVC, the latter with sustainability. That leaves intact
that the scientific credentials of both of them rest upon their long-term
fruitfulness in generating results that are grounded soundly in empirical
evidence. To adopt one of them does not provide a ground to contest
knowledge claims soundly established under the other. Under both of
them, research can produce understanding of phenomena of the world and
their possibilities—and aims to do so, as well as to gain understanding per-
tinent to value-laden interests in application: "political determinants enter
at the point when *basic* [my italics] scientific questions are asked and not
only at the time when technologies are delivered to society."[21]

The competition between AES and BTS, rooted in contested social values, concerns the kind of scientific knowledge that should inform practical
applications, and thus, it also concerns research priorities. Where the values of the market and of MVC are contested, for example, among those
who hold the values of sustainability, there remains no objection in principle to engaging in research under strategies (e.g., AES) which, if fruitful,
can be expected to inform practices that will further the social embodi-

ment of these values. And current fruitfulness suggests that its limits have not been reached. The practices that express the values of sustainability cut across the grain of the global-market project, and in these days of market triumphalism, alternate possibilities are easily discounted. Nevertheless, as Altieri (1995) has documented, numerous groups of small-scale farmers throughout the impoverished regions of the world have made great improvements in their lives and communities through implementations of agroecology, which has become an essential part of their struggle to maintain and develop their cultural heritage as well as to meet their material needs.

Matters for Further Investigation

In this chapter I have not attempted to draw out the far-reaching implications of my conclusion that what seeds are is partly a function of the SCN of which they are constituents. To do so would require a deeper analysis of the meaning of "SCN" than that offered in my brief sketches of the SCN linked with seeds as regenerative resources and as commodities. The sketches suffice for my present purposes: to challenge the argument for the universal value of TI-agriculture, to provide some philosophical legitimation for the AE practices being used by small farmers in many Third World countries, and to urge the importance of conducting more scientific research to support these practices. Once space has been legitimated for alternative agricultural and related research practices, the character of seeds as sociocultural entities (another item on the list of what seeds may be simultaneously) comes to the forefront.

The SCN of seeds may be considered more or less narrowly. "SCN" refers to the whole range of social and cultural relations implicated in the processes and practices—of generation, selection, cultivation, production, processing, distribution, preservation, and consumption—in which the seeds and their products (crops) are used directly. In my sketches I alternated between a narrow and a broad focus: the local community of small farmers when considering agroecology and, when considering transgenics, the vast network of market and property relations often of global reach and impact on an enormous variety of ecological systems. This can be misleading, perhaps suggesting that we confront a stark choice everywhere between the agroecology of small farmers and large-scale high-intensive (progressively becoming TI-intensive) agriculture, or that sustainability defines the only possibilities of interest (agriculture also needs to produce food to feed and nourish the populations of large cities, and often it must provide export crops). With a subtler and more complex concept of SCN, such over-simplification can be avoided, and other dimensions that I have ignored here can be explored. For example, Shiva has argued for affinities of ecofeminism with many forms of agroecology that are developments of traditional practices and knowledge. With this, we can explore the mechanisms by which seeds become resocialized from one kind of sociocultural entity to another, the possibilities for transformation from current predominant agricultural practices, and whether the possibilities of agroecology mentioned above are essentially confined (if they are viable at all) to small and perhaps precarious (Lacey 2000) niches within a larger SCN (shaped by the global market) or can be the basis for a profound transformation of agriculture (at least in some third world countries) that is consistent with the objectives of the many movements throughout the world organizing against the current dominance of the global market.[22]

Conclusion

The transgenic seeds that have been developed and those being developed today are objects of value, not universally, but only within the SCN of their development and application, that in which MVC is deeply embodied and nourished largely by the institutions and structures of the global market.[23] These seeds, with few exceptions, can have no place in AE practices. Conversely, seeds qua the regenerative resources that they are in agroecology have at most a marginal place in TI agriculture. The value of research conducted under AES (and the questioning of the universal value of transgenics) derives not from anti-science sentiments, but from challenging the powerful links of mainstream science (that conducted virtually exclusively under MS) with currently hegemonic values, and from solidarity with poor people whose movements are struggling to recover and enhance their personal and communal agency.

What the seeds used in agriculture are is inseparable from the SCN in which they are planted. It follows that the degree of (social and moral, not epistemic) value accorded the knowledge that informs the use of the different kinds of seeds will be much greater in the SCN of which particular seeds are constituents. What knowledge as well as what practical applications we value most, and thus what strategies we will adopt in research, depend on the SCN in which we attempt to carry on our lives. Conflicts about seeds are an integral part of conflicts about what SCNs are viable and, thus, cannot ultimately be settled without engaging in political struggle. Nor can they be settled without systematic empirical investigation (which requires the adoption of a variety of strategies) of proposals like those that frame the argument for the universal value of transgenics.

Notes

1. V. Shiva, *The Violence of the Green Revolution* (London: Zed Books, 1991); *Biopiracy: The Plunder of Nature and Knowledge* (Boston: South End Press, 1997); hereafter, both are cited in text.

2. The mechanisms of the transformation have been well described by others: J. Kloppenburg, Jr., *First the Seed* (Cambridge, U.K.: Cambridge University Press, 1988); R. Lewontin, "The maturing of capitalist agriculture," *Monthly Review* 50 (3) (1998): 72–84; Shiva (1991, 1997).

3. Material in the previous paragraphs, and in a few places below, has been adapted from H. Lacey and M.B. Oliveira, Preface to V. Shiva, *Biopirataria* [Portuguese translation of Shiva, 1997] (Petrópolis: Editora Vozes, 2001).

4. I defend this assertion in H. Lacey, "Ethics, Agro-Industrial Production and the Environment," paper presented to International Seminar on Ethics and Politics, organized by SESC (Serviço Social do Comércio), São Paulo, Brazil, October 17, 2001; hereafter cited in text as Lacey (2001b)

5. Cf. R. Lewontin, *Biology as Ideology* (New York: HarperCollins, 1992); hereafter cited in text.

6. H. Lacey, *Is Science Value Free?* (London & New York: Routledge, 1999); hereafter cited in text.

7. H. Lacey, "Incommensurability and 'multicultural science,'" in *Incommensurability and Related Matters*, ed. P. Hoyningen-Huene & H. Sankey. (Dordrecht: Kluwer, 2001); hereafter cited in text as Lacey (2001a).

8. D. Tilman, "The greening of the green revolution," *Nature* 396 (1998): 211–2; hereafter cited in text.

9. Also, D. Tilman, "Causes, consequences and ethics of biodiversity," *Nature* 405 (2000): 208–11.

10. Zhu, Y., et al., "Genetic diversity and disease control in rice," *Nature* 406 (2000): 718–22.

11. For the details, see R. Lewontin and J.-P. Berlan, "The political economy of agricultural research: The case of hybrid corn," in *Agroecology*, ed. C.R. Carroll, J.H. Vandemeer, and P.M. Rosset. (New York: McGraw-Hill, 1990).

12. Also: M. Altieri, *Agroecology*, 2nd ed. (Boulder, Colo. Westview, 1995); hereafter cited in the text.

13. See also M. Altieri, "No: Poor farmers won't reap the benefits," *Foreign Policy* 119 (2000): 123–7.

14. H. Lacey, "Seeds and the knowledge they embody," *Peace Review* 12 (2000): 563–69; hereafter cited in the text.

15. My account of *agroecology* is derived mainly from the writings of Altieri (especially Altieri 1995). The term "strategy" is my own (Lacey 1999). A more extensive account is in H. Lacey, "The social location of scientific practices" (paper in preparation). AES are not intended as a full-scale or unique substitute for MS; rather they complement MS (and specifically BTS) and draw upon results obtained under the latter in various ways; for details, see Lacey (2001a). Pointing

to the value of AES is linked with urging that a multiplicity of strategies be encouraged in scientific institutions (Lacey 1999, ch. 10).

16. M. Altieri, *Agroecology* (Boulder, Colo.: Westview, 1987), xiv–xv; hereafter cited in the text.

17. M. Altieri, et al., "Applying agroecology to improve peasant farming systems in Latin America: an impact assessment of NGO strategies," in *Getting Down to Earth: Practical Applications of Ecological Economics,* ed. R. Costanza, et al. (Washington, D.C.: Island Press, 1996), 367–8.

18. See especially J. Kloppenberg, Jr., "The plant germplasm controversy," *Bioscience* 37 (1987): 190–8.

19. M. McGloughlin, "Ten reasons why biotechnology will be important to the developing world," *AgBioForum* 2 (1999): 163–74.

20. Borlaug, N.E. "Ending world hunger: The promise of biotechnology and the threat of antiscience zealotry," *Plant Physiology* 124 (2000): 487–90.

21. M. Altieri, *Biodiversity and Pest Management in Agroecosystems* (New York: The Haworth Press, 1994), 150–1.

22. This section was added in response to particularly insightful comments of an anonymous referee. Throughout I am grateful for helpful comments from Marcos Barbosa de Oliveira and Anna Carolina Regner.

23. I argue elsewhere (Lacey 2001b) that the celebrated recent development of lines of "golden rice"—rice that contains provitamin A in its endosperm, said to be motivated by the humanitarian interest of addressing malnutrition in some third world countries—does not seriously challenge this claim.

References

Altieri, M. *Agroecology: The Scientific Basis of Alternative Agricultures.* Boulder, Colo.: Westview, 1987.

———. *Biodiversity and Pest Management in Agroecosystems.* New York: The Haworth Press, 1994.

———. *Agroecology: The Science of Sustainable Agriculture.* 2nd ed. Boulder, Colo.: Westview, 1995.

———. "No: Poor Farmers Won't Reap the Benefits." *Foreign Policy* 119 (2000): 123–7.

Altieri, M., Yurjevic, A., Von der Weid, J.M., and Sanchez, J. "Applying Agroecology to Improve Peasant Farming Systems in Latin America: an Impact Assessment of NGO Strategies." In *Getting Down to Earth: Practical Applications of Ecological Economics,* ed. R. Costanza, O. Segura, and J. Martinez-Alier. Washington, D.C.: Island Press, 1996.

Borlaug, N.E. "Ending World Hunger: The Promise of Biotechnology and the Threat of Antiscience Zealotry." *Plant Physiology* 124 (2000): 487–90.

Kloppenburg, J., Jr. "The Plant Germplasm Controversy," *Bioscience* 37 (1987): 190–8.

———. *First the Seed.* Cambridge, U.K.: Cambridge University Press, 1998.

Lacey, H. *Is Science Value Free? Values and Scientific Understanding.* London and New York: Routledge, 1999.

———. "Seeds and the Knowledge They Embody." *Peace Review* 12 (2000): 563–9.

———. "Incommensurability and 'Multicultural science.'" In *Incommensurability and Related Matters*, ed. Hoyningen-Huene and H. Sankey. Dordrecht: Kluwer, 2001a.

———. "Ethics, Agro-Industrial Production and the Environment." Paper presented to International Seminar on Ethics and Politics, organized by SESC (Serviço Social do Comércio), São Paulo, Brazil, October 17, 2001b.

Lacey, H., and Oliveira, M.B. Preface to V. Shiva, *Biopirataria*. Petrópolis: Editora Vozes, 2001.

Lewontin, R. *Biology As Ideology*. New York: HarperCollins, 1992.

———. "The Maturing of Capitalist Agriculture." *Monthly Review* 50, (3) (1998): 72–84.

Lewontin, R., and Berlan, J.-P. "The Political Economy of Agricultural Research: The Case of Hybrid Corn." In *Agroecology*, ed. C. R. Carroll, J.H. Vandemeer and P.M. Rosset. New York: McGraw-Hill, 1990.

McGloughlin, M. "Ten Reasons Why Biotechnology Will Be Important to the Developing World." *AgBioForum* 2 (1999): 163–174; ⟨http://www.agbioforum.org⟩.

Shiva, V. *The Violence of the Green Revolution*. London: Zed Books. 1991.

———. *Biopiracy: The Plunder of Nature and Knowledge*. Boston: South End Press, 1997.

Tilman, D. "The Greening of the Green Revolution." *Nature* 396 (1998): 211–2.

———. "Causes, Consequences and Ethics of Biodiversity." *Nature* 405 (2000): 208–11.

Zhu, Y., et al. "Genetic Diversity and Disease Control in Rice." *Nature*, 406 (2000): 718–22.

Fallout: Issues in the Study, Treatment, and Reparations of Exposed Marshall Islanders

Robert P. Crease

I want to begin with an incident that took place on Bikini in the Marshall Islands shortly after World War II. A young U.S. lieutenant who had not been to the South Pacific before was put in charge of cleaning up the island and specifically ordered to preserve the local flora and fauna as much as possible. He had the refuse swept into neat piles and decided to burn it as the most efficient means of disposal. Unfortunately, nobody had told him about burning coconuts. When coconuts burn they explode, with a force that hurls flaming bits of husk dozens of feet. The lieutenant's carefully separated piles thus spread the fires to the surroundings, soon creating a conflagration that consumed the very local flora and fauna he had been ordered to protect.[1]

This incident symbolizes much of what I want to point out about the following story. The story involves Western interventions in non-Western cultures in politically volatile situations. It includes a scientific-technical dimension in which people have lost confidence in traditional sources of authority. It concerns the actions of U.S. doctors, politicians, and activists seeking to aid Marshallese inhabitants exposed to fallout in the wake of a nuclear weapons test. Thanks to unfamiliarity with the culture and environment, actions taken with the best of intentions created incendiary situations, which wound up harming the very people the actions were designed to assist. The story raises a number of philosophical issues, including the role of research versus therapy in medicine, the nature and ethics of speaking for the "other," and the ethics of experimentation with socially vulnerable people. The story is so unique as to create danger of shoe-horning it into traditional social movement narratives of victimization or oppression—the civil rights struggle, the struggle against cultural imperialism, the Tuskegee syphilis experiments, and so on—and allowing those narratives to define the issues. The Marshallese story belongs to a different genre. I cannot hope to explore in depth the issues, or even the extended story. What I'll try to do is tell enough of the story to exhibit these issues and suggest their larger significance.

Let me state right away my prejudices and approach. I'm the historian at Brookhaven National Laboratory, which from 1956 to 1998 was one of the player institutions, and I'm also a professor at the State University of New York, Stony Brook, which now participates in Brookhaven's management. This position situates me by exposing me strongly to the perspective of those within that institution—I accompanied one of the Brookhaven team's last research trips to the Marshall Islands—and by giving me unprecedented resources such as open access to Brookhaven documents related to the program. It also means I begin with a perspective that apprehends certain distinctions as significant that other participants may not, especially that between Brookhaven and other player institutions like the U.S. Atomic Energy Commission (AEC). But I shall adopt a hermeneutical approach, which provides for the possibility of transforming situated perspectives without escaping from them. I can characterize my use of that approach the way Clifford Geertz does at one point in his recent book, by saying that it begins by seeking to discover who the actors think they are, what they think they are doing, and to what end they think they are doing it (Geertz 2000, 16).[2] This does not, of course, entitle us to speak for the actors. But it does create the groundwork for critique; it does not lead to descriptions which would suspend judgment, nor to the application of ideologies which would suspend reflection. Thus I agree with anticolonialist historian Klaus Neumann that what's called for is not adding new voices to the old narrative, or "empathizing with the oppressed other and focusing on anti-colonial resistance" (Neumann 2000, 27), for such a path easily devolves into patronizing and even infantilizing talk and behavior.

The Marshall Islands would seem a highly unlikely place for a study of science and society. Located in the Central Pacific, they consist of two chains of atolls whose total area is about 700 square miles. The main staple is fish and fruit, and the most popular delicacy, the coconut crab. For about a century and a half the history of the Marshalls was colonial. They were named for a commanding officer of a British ship; they became a German protectorate in the 1880s; they were occupied by Japan in World War I (with a League of Nations mandate to administer them), which fortified them and used them as military bases. In 1944, the United States captured the islands from the Japanese, and after the war, the United Nations Security Council designated the United States as the administrator of the Trust Territory (TT) of the Pacific Islands. That agreement obligated the United States to "protect the health of the inhabitants" and encourage their self-sufficiency, but also entitled the United States to establish military bases, erect fortifications, and close off areas for security purposes.[3] The United States used the Bikini and Enewetak atolls in the western chain of the Marshalls in testing nuclear weapons and resettled the inhabitants.

On March 1, 1954, the first "dry" or potentially deliverable thermonuclear bomb (one not requiring a refrigerator) was detonated as part of the testing program on Bikini. It was code-named "Bravo." While Bravo had been estimated to be more powerful than previous blasts, the sponsoring agencies—the Atomic Energy Commission (AEC) and the Defense Department (DD)—did not plan the same kind of elaborate precautions to protect the Marshallese inhabitants as before. One reason was that little local fallout had accompanied the previous test, "Mike," and little danger was foreseen from Bravo; in the ten to fifteen hours it would take radioactive debris to reach the closest inhabited areas in the eastern chain (which included the Rongelap, Rongerik, Ailingnae, and Utirik atolls), much of the material would have decayed. Another reason was monetary. Reductions in AEC service budgets for 1954 led officials to stress the need for austerity, and evacuations were expensive. Officials felt that if evacuations were needed, they could be carried out by ships of Joint Task Force Seven (JTF), whose main duty would be to chase away Russian submarines. As for radiation dose, a double standard was applied; the maximum permissible exposure (MPE) for task force personnel was 3.9 R (0.3 R a week for thirteen weeks), with waivers possible up to 20 R, while 20 R was considered acceptable for native populations without waiver.

The firing time depended mainly on the weather, which appeared favorable until a few hours before the test, when it was viewed as unfavorable because of winds that might blow the fallout towards inhabited areas. But at the measured speeds and altitudes there still seemed no danger of unsafe exposure to inhabited areas, and Bravo went off on schedule (Hacker 1994).

In military terms, Bravo was wildly successful, with a four-mile-wide fireball and a fifteen megaton yield, three times greater than expected and nearly a thousand times the force of the bombs that had destroyed Hiroshima and Nagasaki nine years before. But far more fallout had been produced than anticipated. The JTF-7 ships fled the area at top speed, their surfaces covered by a white radioactive dust. Instrument problems, confusing messages, and delayed data reception misled test officials into thinking that populated areas were not in danger. By midnight, it was clear that inhabited areas had in fact been exposed to high levels of radiation. The next day, March 2, military personnel were evacuated from a weather station at Rongerik, 135 miles east of Bikini, and evacuations were ordered for the native populations of that atoll and others nearby. On March 3–4, the destroyer escorts were sent to pick up inhabitants on Rongelap (64 people), Ailingnae (18), and Utirik (154) and transfer them to Kwajalein Island, the largest island in the eastern chain and a U.S. military base. Some had been exposed to radiation for over two days.[4]

An emergency medical research team, jointly sponsored by the AEC and the DOD, was quickly dispatched to the area, reaching Kwajalein on March 8. Its mission was to study the effects of the exposure to fallout and carry out needed treatment. The twenty-one-person team was headed by Eugene Cronkite, then of the Naval Medical Research Institute.

Project 4.1, as Cronkite's team came to be called, arrived at Kwajalein on March 8. They found the place in utter confusion, with temporary shelters put up to house the evacuees. Medically, too, the situation was confusing, for the AEC had collected no information on the medical histories of the population, and had installed no dosimeters on the islands to record dose information. In view of later insinuations and accusations that the exposure had been part of a deliberate test, this is the clearest indication to the contrary. In all previous AEC radiation experiments, an extraordinary amount of preparatory work was involved, in the absence of which any data collected would be worthless (Hacker 1994).

Here I want to point to a first issue, which I'll call *narrative shoe-horning*. Documents relating to the firing-time decision, including memos, charts, and weather data, have been carefully reviewed by historians and investigatory panels (Hacker 1994; Hewlett and Holl, 1989; Advisory Committee 1996). No evidence emerged that the exposure was deliberate. Nevertheless, many Marshallese continue to believe that it was, and it has become all but a factoid in popular culture (see, for example, the best-selling novel *The Web*, Kellerman 1995). The supposed deliberate character of the exposure has become a symbol of Western medical exploitation of nonwhite races—a nuclear Tuskegee. There are many reasons why belief that the exposure was deliberate is so widespread, pervasive, and persistent—some articulated by the Advisory Committee after hearing presentations by representatives of the Marshall Islands.[5] But another is surely that this charge has become a useful reference point for the moral claims of Marshallese for increased U.S. medical and economic reparations. It has become, in short, a crucial element in what Benford calls a "movement narrative." Benford argues, building on research into the role of narratives in meaning formation and applying it specifically to social movements, that narratives "seem particularly relevant to movements considering their centrality to the social construction, maintenance, and diffusion of vocabularies of motive, collective identities, collective memories, collective action frames, and affect." Movement narratives help shape Collective Action Frames, which "punctuate or single out some existing social action or aspect of life and define it as unjust, intolerable, and deserving of corrective action" (Snow and Benford 1992, 137). Movement narratives thus "function as internal social control mechanisms, channeling and constraining individual as well as collective sentiments, emotions, and action" (Benford 2002), and must be promoted and reinforced lest alternative narratives be constructed and disseminated.

In the case of the Marshallese exposure, the suggestion of deliberate intent is often promoted by reference to the exposure as a "so-called" or "alleged" accident—conspicuously surrounding the adjective with scare quotes—with the corollary that only a continuing conspiracy to conceal evidence prevents us from characterizing the exposure as deliberate. But this suggestion ultimately threatens to sidetrack and misrepresent the key issues. For the storytellers, the conspiracy theory is self-serving because it provides false moral consolation. It shifts the blame onto a specific set of conspirators, the anomalous Americans—immoral madmen—in charge of the Bravo test: we, after all, weren't among them. The moral situation is murkier the other way. And by serving to center the case for increased reparations on the deliberate character of the exposure, the conspiracy theory diverts attention from the larger, ongoing, impact of the United States on Marshallese society. Thus the suggestion of conspiracy, intended to reinforce the case for U.S. reparations—and while effective at riveting public attention—ultimately serves to undermine it. For isn't the most morally shocking and commanding aspect of the Tuskegee study the fact that it was *not* a conspiracy?

Back to the story of the exposed Marshallese. The Utirik inhabitants were found to be free of symptoms of radiation sickness, and were returned to their island in June 1954. Those from Rongelap and Ailiginae were less fortunate. They displayed typical symptoms of radiation exposure; vomiting, nausea, diarrhea, low blood counts, temporary hair loss, and skin lesions. For several years they would have a slight increase in the rate of stillbirths and miscarriages. Furthermore, their islands were deemed uninhabitable, and they were transferred to Ejit Island in another atoll. The

Rongelapese wanted to return to their native island as soon as possible, and the AEC faced international pressure to do so; in 1957, when the residual radiation on the island fell into what was considered a safe level, the islanders were resettled.[6]

In order to provide continuing care for the exposed Marshallese, an arrangement was worked out between the Trust Territory and the AEC. The normal medical care for the Marshallese—including the ordinary medical needs of the exposed population—would be the responsibility of the Trust Territory. The exposure-related studies and care of the exposure victims would be the responsibility of the AEC, through its Division of Biology and Medicine (DBM), which planned six- and twelve-month follow-up visits.[7] ("Responsibilities for Care and Disposition of Native Inhabitants of Rongelap and Utirik Atolls," Joint Task Force Seven, July 6, 1954.)

But the AEC was unhappy with this burden, and in 1956 subcontracted Brookhaven National Laboratory to take over responsibility for the exposure-related studies and care of the Marshallese. Brookhaven was an obvious choice. In the two years since the exposure, several members of the Project 4.1 team had left the Navy and joined the lab, including project head E.P. Cronkite, Robert Conard (who had also been at the Naval Medical Research Institute), and Victor Bond of the Naval Radiological Defense Laboratory. But Brookhaven was also a multidisciplinary laboratory whose life sciences departments were outfitted to study radiation and its effects in a variety of contexts. One of Brookhaven's first duties, at the AEC's request, was to compile a scholarly report on the exposure, the so-called Green Book, the first major study of effects of fallout on human beings (Cronkite et al. 1956).[8]

At the time Brookhaven took over responsibility for these surveys, no fatalities had occurred among the Marshallese, and the effects of fallout seemed to have all but disappeared. "All of the exposed individuals have recovered from the immediate effects without serious sequelae," reads the Green Book. Conard, head of the new Brookhaven study team, reassured islanders that the danger had passed. Studies of Japanese atomic-attack survivors, however, had revealed an increase in susceptibility to cancer, and the number of people exposed to the doses evidently received by the Marshallese was too small to make predictions. The Brookhaven team drew up plans for long-term medical care and study of the victims, recommending that annual examinations be continued indefinitely. Thus began Brookhaven's forty-two-year commitment, under contract initially with the AEC, later with the Department of Energy, to examination of the Marshallese. The commitment was to involve annual visits, treating the radiation-related illnesses of the Marshallese, conducting procedures which consisted mainly of taking blood and urine samples and measuring radiation levels, and publication of the results of the study in medical literature.

This raises a second issue—I'll call it the *guinea pig charge*—involving research with human subjects. The Brookhaven team's mandate was two-fold: to provide medical treatment for the exposure victims, and to study the effects of the exposure. This study in turn had a dual function: to guide the medical treatment of the Marshallese, and to allow comparison with other victims of radiation exposure in a way that might help guide treatment for victims of potential future accidents. The knowledge, in short, would be simultaneously for the victims and a broader community via the scientific literature.

I do not want to broach, yet, the question of the significance of the sharp cultural difference between the doctors and patients, which uneasily resembles the traditional split, in colonial societies, between "those who know and decide and those who are known and are decided for" (Geertz 95), and which can foster situations in which the burdens and benefits of the research are inequitably distributed and human subjects used merely as means for furthering the interests of the researchers. My point here is just the familiar and obvious one that there's a spectrum of possible ways in which researchers can position themselves with respect to human subjects. All involve treating the human subjects as means to some extent, but it only becomes mere means, and hence unethical by our conventional definitions, when the researchers' focus is not on the overall situation of the human subjects but solely on the research results. Only here does the irresistible and incendiary sobriquet "guinea pigs" become anywhere near appropriate.

And from the perspective of the U.S. authorities, the AEC, and Brookhaven doctors, this was not the case. The Brookhaven studies were retrospective, follow-up studies in the wake of a terrible accident, analogous in their eyes to studies any responsible doctor would have ordered on an accident victim, and it would have been unethical (medical malpractice) not to have conducted them. That the results were sufficiently instructive and of interest to warrant publication in scientific literature did not ipso facto imply the researchers were using their subjects as mere means. The medical prognosis of the unfortunate Marshallese mattered to them and to humankind at large, for it helped Brookhaven's medical department formulate for the first time a comprehensive program for treatment of radiation victims. As Barton Hacker, the former laboratory historian at Lawrence Livermore National Laboratory, wrote, "Like the American radium-dial painters of the 1920s and the Japanese of Hiroshima and Nagasaki in 1945, the Marshallese of 1954 inadvertently were to provide otherwise unobtain-able data on the human consequences of high radiation exposures" (Hacker 1994). Also, didn't the Brookhaven doctors—and team head Conard in particular—regard themselves as the natives' friend, cultivating and valu-ing their Marshallese friendships? Didn't the Brookhaven doctors explain the reasons for their tests before asking permission to conduct them? Didn't they only carry out tests they would have performed on themselves? Hadn't Conard even collected a representative sample of the very slightly radioac-tive Rongelapese food, consumed it, and measured its effects on his own body, in order to carry out a controlled study?[9] From this perspective, to accuse the Brookhaven doctors of experimenting *on* the Marshallese was a category mistake.

As noted by the Advisory Committee on Human Radiation Experi-ments, the AEC, the JTF, and the Brookhaven doctors all expressed enor-mous interest in the research on radiation effects that the exposure made possible—but it was also noted this did not mean that such research was the only goal, nor did it make the research unethical. The Advisory Com-mittee also reviewed the work of the Brookhaven team specifically to look for any tests that did not serve primarily to benefit the subjects and found two, but they also found these were not harmful to the subjects and were done in order to learn about radiation effects in general (Advisory Com-mittee 1996, 373).[10] In fact, "the Committee believes that the AEC had an ethical imperative to take advantage of the unique opportunity posed by the fallout from Bravo to learn as much as possible about radiation effects in humans" (Advisory Committee 1996, 376–7).

But now let's look at the situation from the perspective of the Mar-shallese. Westerners may witness a close interaction between the specialists who study and the physicians who treat them, but to the Marshallese, there appeared to be no connection between the studies in which they were sub-jects and the rest of their medical care. Recall the division of labor: the local care was supposedly the responsibility of the Trust Territory, while the expo-sure-related study and treatment was provided by a completely different source, the Atomic Energy Commission, subcontracted to Brookhaven Na-tional Laboratory. The one was woefully inadequate, mainly consisting of a physician's assistant equipped with some rudimentary medical supplies and a short-wave radio able to reach Kwajalein, while the other—the one tied to the research ambition—was plentifully supported and consisted of first-rate doctors equipped with the latest devices and skilled in the most advanced procedures. What signal was being sent about which had priority?

Also, the interactions between the Brookhaven team and the Marshallese in the 1950s and 1960s, while generally friendly, were accompanied by little of what Geertz calls "deep hanging out." Here's a description by Brookhaven's resident physician in the islands from 1975 to 1976, attempting to describe what the examinations were like from the Marshallese point of view:

> [E]ach March a large white ship arrives at your island. Doctors step ashore, lists in hand of things to do, and people to see. Each day a jeep goes out to collect people for examinations, totally interrupting a normal daily activities. Each person is given a routing slip which is checked off when things are done. They are interviewed by a Marshallese, then examined by a white doctor who does not speak their language and usually without the benefit of a Marshallese man or woman interpreter. Their blood is taken, they are measured, and at times, subjected to body scans.[11]

No members of the Brookhaven team learned the Marshallese language. When they came, they lived on shipboard and ate food they had brought along. The visits were run on a Western schedule. Conard, under pressure from other doctors on his team who needed to get back to their practices, would compress the schedule as much as possible and often tried to examine on Sundays. He would approach the magistrate and ask permission, the magistrate would nod his head, and, not entirely familiar with the intricacies of politeness in the Marshall Islands, Conard would go away thinking, erroneously, that everything was arranged.

The lack of cross-cultural understanding is exemplified by the matter of the coconut crabs. Many residual radioisotopes on Rongelap, such as cesium 137, either remained in the ground or cycled between coconut trees and ground without concentrating enough to pose a danger to humans. Strontium 90 did, too—except for one species, the coconut crab. This crab, whose formidable jaws are strong enough to crack open coconuts, lives on coconut milk. Coconut milk is high in calcium and therefore also concentrates the chemically similar strontium, directly beneath calcium on the periodic table. Coconut crabs therefore concentrate strontium, especially in the rich meaty part which is the Marshallese delicacy. But when Conard first told the Marshallese they were forbidden to eat the crabs, he was initially confused by the difference between coconut and land crabs and then was puzzled as to why his injunction was not fully obeyed. "They apparently use them for food to a greater extent than I had supposed," he wrote.[12]

At this point, one begins to guess that the interesting question is not why the Brookhaven scientists were about to lose their authority, but rather what had created and maintained it in the first place. The answer can be explained via something akin to Bordieu's image of society as a structured social field in which social agents take up allowable positions whose possibilities for action are defined by their political "capital." The position occupied by the Brookhaven scientists in that field—their ability to show up year after year, actually on the beach, and conduct complete physical examinations of the inhabitants—was a legacy of the traditional acquiescence of Marshallese leaders to colonial powers.

As the 1960s wore on, that field would be restructured. An important influence was the growing antinuclear weapons movement in Japan; activists there from the Japan Congress Against Atomic and Hydrogen Bombs (the Gensuikin) began to make contact with Marshallese politicians, seeking common cause, and invited many to their gatherings. Marshallese politicians, one must add, wield enormous influence. In a region where land is extremely scarce, landowners tend to accumulate extraordinary wealth and power even as much of the population remains at the subsistence level. The Marshallese politicians then began to adopt the movement narrative of the Japanese atomic survivors for their own cause. The subsequent restructuring of the Marshallese political field inevitably changed the position of the Brookhaven doctors in it. The result was a situation highly vulnerable to exploding coconuts. And meanwhile, several coconuts had been tossed, with good intent, into the fire.

One was that the Brookhaven team, noting the hopeless inadequacy of the local health care, attempted to fill the void during its visits by increasing its routine medical care to the extent possible within the time and fiscal limitations. Islanders not exposed to fallout were included in the examinations, immunizations were provided that the Trust Territory should have been providing but were not, and the Brookhaven team began administering to non-radiation-related health problems. A dentist, for instance, which the Islanders had never had, was included though extracted teeth were then examined for traces of strontium. This was technically illegal, because the Congressional authorization did not permit funds for ordinary health care to come out of the AEC budget that supported the Brookhaven project. The result was to blur the distinction between the research activity and the activity of providing ordinary health care needs. The gap remained wide between the needs of the inhabitants and the treatment the Brookhaven team could provide—but the expectation had been created among the Marshallese that the hope and even responsibility for filling the gap was up to Brookhaven.

Another coconut consisted of U.S. Peace Corps volunteers. Anticipating the eventual end of Trust Territory administration and the beginning of negotiations to replace it with a treaty, the United States began actively seeking ways to favorably dispose Marshallese public opinion. One bright idea was to send in Peace Corps volunteers. But remember this is the late 1960s, when young Americans attracted to the Peace Corps tended not to be favorably disposed to the Nixon administration, nor to U.S. attempts to impose itself abroad and especially in non-Western countries. They were only too willing to encourage the thus-far relatively quiescent Marshallese to voice grievances, protest abuses, and challenge U.S. hegemony.

Finally, there was an unexpected medical development. Beginning in the mid-1960s, many exposed Marshallese began to develop thyroid abnormalities including reduced function, tumors benign and malignant, and stunted growth in children. Previously, the principal source of danger to the thyroid gland from fallout was thought to be iodine 131. The Brookhaven medical team now realized that several other extremely short-lived iodine isotopes had been present, which also had harmful long-term effects, and that a range of thyroid abnormalities besides cancer could result. Many Marshallese required thyroid operations, which were carried out in the United States, beginning in 1965. From then on, Marshallese were brought almost yearly to the United States, first to Brookhaven for extensive workups, then to hospitals in Boston or Cleveland for the operation.

In the 1970s, these coconuts began to explode. Marshallese politician Ataji Balos, one of those whom the Gensuikin had invited to Japan, invited a fact-finding team, mainly of Japanese antinuclear activists, to visit the island, but when the team arrived in December it was refused admission to visit Rongelap by Trust Territory authorities, on the grounds that its members had only tourist visas. Marshallese politicians were outraged, and anti-U.S. language mounted. From their point of view, distinctions between Brookhaven, the AEC, the DD, and the Trust Territory—who was responsible for what—meant next to nothing; they all belonged to the same system. In January 1972, Balos charged that the United States was using the Marshallese as guinea pigs and had deliberately exposed the brown-skinned natives to the fallout. He said the principal mission of the Brookhaven team was to collect medical data so that it could develop a treatment program for those who might be exposed during wartime. This last remark, of course, was half correct.

In March, when the Brookhaven team arrived for its annual checkup, it discovered that Balos had told the inhabitants to refuse to participate, and the examinations could not be carried out. Conard was shocked and puzzled that his personal and scientific appeals, so effective in the past, now had no impact. One Rongelapese told a reporter on the scene, "We are very sorry we're not allowed to see the doctors. I am not sure why but we are afraid of our leaders."[13]

That fall, following discussions between AEC officials and Marshallese politicians, the Brookhaven team was allowed to return. They discovered that a nineteen-year-old male, Lekoj Anjain, son of John Anjain, magistrate of Rongelap at the time of the Bravo test, had an acute case of leukemia, almost certainly related to his radiation exposure as an infant. Lekoj was flown to Brookhaven where his diagnosis was confirmed. He was moved to the National Institutes of Health, where he shared a hospital room with columnist Stewart Alsop. In November, Lekoj died.[14]

There followed a bitter fiasco. Lekoj's mother and father came to escort the body back to Rongelap. They flew from New York to Honolulu, but mechanical difficulties forced a change of planes in Los Angeles, and in the process Continental Airlines misplaced the coffin. The distraught mother and father had to return to the Marshall Islands without it. The coffin finally arrived at Kwajalein, and services were held in nearby Ebeye before a plane was to take it to Rongelap. The plane ripped its bottom out inside the lagoon and became unavailable, and a Trust Territory vessel was pressed into service. After the funeral, the vessel set out to return to Rongelap, carrying two hundred passengers from Ebeye who had attended the Rongelap services. The night was foggy, and at 2 A.M. the boat, traveling at full speed, plowed into a reef at high tide and could not be dislodged. The frightened passengers huddled on board until daylight, when they observed a small island a few hundred yards distant, where they were ferried until retrieved by a passing commercial boat.

The fatality, the first among the Marshallese probably due to the exposure, further inflamed the climate. Charges were again lodged that the United States had deliberately exposed the islanders. Conard's reassuring statements that the danger had passed, made before discovery of the thyroid abnormalities, were recalled as evidence of the conspiracy. In the wake of Lekoj's death, the Brookhaven program was extended to twice a year, and the AEC agreed to allow Brookhaven to station a resident physician year-round on the islands.

The Peace Corps coconuts then exploded. The volunteers, many of whom—unlike the Brookhaven doctors—learned the language, ate the food, and lived on the islands, quickly took up influential positions in the social field, exactly as hoped by the U.S. authorities. But instead of promoting pro-American loyalty, they sometimes actively encouraged the Marshallese to demand better medical treatment than the shoddy and inadequate services they were being given. Some openly accused the Brookhaven doctors of negligence, of incompetence, of perpetrating abuses, and of being more interested in the scientific findings than the health of the Marshallese.

The Brookhaven doctors found all this baffling and disturbing. They saw themselves as serious medical professionals on a humanitarian mission, at the request of the U.S. government, serving the best interests of the Marshallese. In a moving personal letter, one wrote:

> To him [Balos] and his clients we represented AEC, and AEC was running some sinister experiment with Marshallese objects. I thought the idea was too ludicrous to be taken seriously, and there I was wrong. The thoughts are taken seriously where it counts most—among the Rongelap people. I hoped that by work and personal acquaintance I could make them see us in a different light and cooperate with us in their own interest.[15]

But in the current atmosphere, he continued, this is impossible. "I am not optimistic about the future of the program."

To better understand this situation, I'd like to compare it with that presented in the Ibsen play, *Enemy of the People*. The central character, Thomas Stockmann, is medical officer at a small spa on which the livelihood of his town depends. He discovers an invisible poison is polluting the spa's water. Wanting to be sure, he sends samples to an outsider—a chemist at the University—for chemical analysis. The results, which he receives in Act One, prove the existence of a toxin in the water, which has sickened many people and will cause far worse harm in the future. He sets out to tell this to the community, thinking they will welcome the news, raise his salary, thank him, even celebrate him.

But it's an exploding coconut story. In Act Two his friend warns him not to be so sure the community will treat him as a hero: "You're a doctor and a man of science, and to you this business of the water is something to be considered in isolation. I think you don't perhaps realize how it's tied up with a lot of other things." Stockmann, for whom the technical scientific data is authoritative, replies, "I don't quite understand you," and insists that the "shrewd and intelligent" people will be "forced" to accept the news. The mayor tells Stockmann that the matter is as much an economic as a scientific one, and the two trade charges over who is arrogant and who has the real interests of town at heart. The local newspaper, the *People's Messenger*, refuses to publish his side of the story, the politicians fire him, and the locals condemn him as an "enemy of the people," regarding themselves thereby as "progressive voices" who speak for the little guy. He's accused of conspiracy because he secretly conducted tests, and the people turn on him violently and stone his house. There are differences between this situation and the Marshallese. But the basic conflict is strikingly similar—a scientist who, of course, accepts a proper technical scientific study as authoritative, on one side, and on the other, a society whose leaders not only don't find technical information authoritative but find it and the scientist threatening to their well-being.

When I've tried invoking Ibsen in talks to scientists, the audience tends to get impatient. What could fictional situations possibly tell us about real life? But think of Ibsen's invented situation as a model that strips away grubby details to present a conflictual situation in a way to make perspicuous its essential forces. One can put questions to it and get answers. And what's clear from Ibsen's model is that this volatile situation with a scientific-technical dimension in which people have lost confidence in traditional sources of authority cannot meaningfully be treated as the product of scientific illiteracy. It is tempting to look upon it as a technical issue, as Stockmann does, and to say, "They just don't understand." But that would be wrong. In Stockmann's town, they are able to understand the threat of bacteria all too well. Nor can the situation be understood as an issue of the media or politics. The media and politicians goad, distort, and complicate, but follow other leads. Finally, the conflict cannot be treated as the product of irrationality. In the Ibsen play, it's all too rational what is happening. The community leaders, whose world is threatened by Stockmann's desire to close the baths, are being eminently practical.

Stockmann, of course, thinks that they are impractical and that he is the practical one. But what is judged practical is not a technical issue but a social and interpretive one. People don't take in technical information nakedly, but interpret it against a background that includes their training, who says the technical information, their experience of how trustworthy that information has been, and so forth. As a result, what looks to a scientist like a technical issue can be to nonscientists a more social issue. Stockmann is able to take technical information, relate it to other pieces of information, integrate through the whole, and see the practical implications. And there is, potentially at least, common social ground between Stockmann and the townspeople. But there is nothing connecting Stockmann's experience of the scientific process and his ability to discourse about it, and that of the townspeople. A huge gap exists between the "load," as it were, born by the discourse in the two cases. To connect the two would require not translation, but a more complicated process, which one might call a discursive "impedance matching." But Stockmann can't even see the mismatch. He's stuck by virtue of habits he can't shake, and he can't even see.

Which is what makes him not a hero—not like some of the women in Ibsen's later plays, who are equally out of place in their own homes. Ibsen leaves us at this stuck position: all Stockmann initially can think of is to flee the town, and then to start all over with children in a hopeless reeducation plan worthy of Plato's *Republic*.

Stockmann belongs to the same world as the townspeople; in fact, he's lived in the town all his life and is connected to its institutions. Imagine Stockmann coming from a Western country with his technical message into a non-Western community with a colonial history. The impedance mismatch becomes much more extreme. It might take far more than one or two instances of deep hanging out.

An episode from the Marshallese story can illustrate. In 1978, Brookhaven finally put a person on Rongelap year-round who spoke Marshallese. Jan Naidu was an ecologist who accompanied the team to help monitor the environment and the dietary exposure. He picked up the language easily because he was Indian and many Marshallese words had similarities with Tamil. He ate the food of the Marshallese, lived with them, fished, and made copra with them. Though Hindu, he attended their church services. And he decided to try to help the Rongelapese understand the science behind their exposure, which was extremely difficult because their language had very few words to explain the required concepts.

So he would sit down in the evening under a coconut tree after eating dinner with them and converse. He used simple terms as much as possible. Energy was easy. When you are sleeping, you don't have much energy. But when you get up, and you are active, your energy goes up, and after you use your energy, you become less active. That's similar to atomic energy; it can have more activity and less activity. Atoms were also relatively easy. Rocks are big and solid; but waves can make them smaller, turning them into fine grains.

When Naidu got stuck, he would ask them for suggestions. He said that when atoms had energy, they could send out some things a long distance and some things a short distance, and he asked them if they knew anything like that. They said, "Aha! Let us compare it to fish! A tuna is large and has a capacity to travel far," like what Naidu told them of one type, gamma radiation, whereas the small fish which cannot keep pace with the tuna can be like other types. He said certain radiations could penetrate if they had a lot of energy, and they said that was like the way you can throw coconut fronds at a house, but a hurricane can throw it through the house.

Naidu held his informal conversations every evening five days a week. They were completely unstructured; some people fell asleep, some left and came back. And they developed tremendous respect for him. He wanted to stay with them another year, but the Department of Energy discontinued the program he was on, ultimately deciding instead to put the funds into producing colorful brochures, in English and Marshallese, to explain radiation to the inhabitants—a program that became a disaster. If the mismatch between the actions of the Brookhaven doctors and the Marshallese had been a question of scientific literacy, what Naidu attempted would have sufficed.

I shall skip events of the next few years, including the filing of a lawsuit on behalf of the Marshallese, the resettlement of the Rongelapese by Greenpeace to a supposedly safer island, and the Compact of Free Association. Fast forward to September 1996, when the Marshallese Nitijela, or Parliament, passed an official resolution declaring that "diabetes and cataracts" were medical conditions "irrefutably presumed to be the result of the Nuclear Testing Program."[16] They ordered the appropriate executive body to issue new regulations covering diabetes and cataracts under the regulations. The action was defended by politicians and ex-Peace Corps workers employed by the politicians.

The causes of diabetes and cataracts have now been extensively studied. There is no evidence linking them with radiation exposure. Is this, then, the worst nightmare of the opponents of social constructivism—that the authority of science has been so undermined that determination of etiology is now up to politicians? The intent, of course, is to unlock more of the U.S. money made available for illnesses to the nuclear testing program— thus, an eminently practical action. And the irony is that the rampant diabetes among the Rongelapese is surely related, not to the radiation exposure caused by the U.S. tests, but to another U.S.-related reason: the dietary and cultural disruptions caused by the U.S. resettlements of the inhabitants and their increasing dependence on canned foods. But the politicians approached the technical information the way lawyers handle evidence in a court case, or advertising copywriters handle data—not as the outcome of a process which may motivate us to reshape our decisions and practical goals, but as a bank of potential weapons to be put in the service of an already decided upon course of action.

Recently I heard Sandra Harding pose the question: Is the philosophy of science silent on issues relating to colonialism and diversity? The fear behind the question, I think, is that if we admit the authority of technical information in certain social arenas, the technical will replace the social and will disable critique: our moral drive will be stilled, our values watered down, our moral dimension lost. But if science is *not* entirely socially constructed, that's precisely what gives it a profoundly political, and potentially critical, force. Science can help navigate in a world of toxins, fear of toxins, litigation over toxins, and those who manipulate fear of toxins. It can help protect against attempts to use others, and socially vulnerable groups in particular, as means for pursuing personal utopian quests. To agree, we need not enter the hopelessly abstract debate sometimes posed in postmodernist science studies of whether all knowledge is a disguised power grab, or whether all science is necessarily tied to the quest for a grand narrative, totalizing pattern, or cosmic blueprint. We need only agree with Geertz that the aim is simply to help create "guidelines for navigating in a splintered, disassembled world."

The day after finishing *An Enemy of the People*, Ibsen wrote a friend that he did not know whether to call his new work a comedy or a drama, for it had both comic aspects and a serious theme.[17] What's comic is the way Stockmann doesn't understand why he's stuck making the same mistakes or even that he is, and how little at home he really is in his native town—an effect underscored by the broken windows in the last act. What's tragic are the consequences of that unhomeliness, which will affect virtually every member of the community. I propose a new category, which might flippantly be called the *exploding coconut* genre. More formally, it might be called the *catastrophe* genre, in the engineering sense of the word; what happens when a complex system grows out of synch with its environment, so that it operates for a time in a domain of instability, until an incident causes the system to break down or operate in a drastically new mode. The story of the Marshall Island exposure victims, I think, embodies another catastrophe story, one that involves a scientific dimension. The catastrophe is brought about not by the technical replacing the social—by science disabling critique and disarming social movements—but just the opposite: by it ceasing to inform practical action, ceasing to provide the basis for critique of a situation. That danger is especially strong in interactions with socially vulnerable populations and colonial environments. The Marshallese catastrophe story raises far-reaching questions about social activism in volatile situations with a technical dimension. Only by paying attention to those questions do we fully do justice to the memory of the exposed victims.

Notes

1. Cronkite, BNL Video Interview.
2. "This," Geertz continues, "does not involve feeling anyone else's feelings, or thinking anyone elses's thoughts . . . nor does it involve going native." It is also considerably more difficult than "the glossy impressionism of pop art 'cultural studies' would suggest."

3. U.S. Department of State, "Trusteeship Agreement," repr. *in Trust Territories of the Pacific Islands,* 1993, Appendix B.

4. The Marshallese and the American personnel were not the only fallout victims. Unbeknownst for two weeks to the military and the public, a Japanese fishing vessel with twenty-three crew members, the Lucky Dragon, had been in the way of the unexpectedly long plume of fallout, which had covered the craft with a white, snow-like substance. Its crew displayed symptoms of radiation sickness. The saga of the Lucky Dragon and the "ashes of death" became a cause celebre, and treatment of the crew became the object of a tug-of-war between American and Japanese officials. The Marshallese victims received much less press attention at the time, but benefitted from superior immediate medical care. Their political exploitation would come later.

5. "There is, of necessity, some tension between data gathering and patient care when the same physician is responsible for both. The Advisory Committee has found no clear-cut instance in which this tension was likely to have caused harm to patients, but some may have been subjected to biomedical tests for the primary purpose of learning more about radiation effects. This inherent tension, coupled with the additional strains of language and cultural differences between the Marshall Islanders and the physicians, appears to have compromised the process of informing the subjects of the purpose of the tests and of obtaining their consent, which has doubtless contributed to their sense of being treated as guinea pigs. Insensitivity to cultural differences, failure to involve the Marshallese in the planning and implementation of the research and medical care program, divided responsibilities for general medical care, and failure to be fully open about hazardous conditions have all contributed to unfortunate and probably avoidable distrust of the American medical program by the Marshallese" (Advisory Committee 1996, 368–9).

6. As Conard wrote in March 1956, "We are committed to return the people to their homes and that is their express wish" (Faden, 1996, 372).

7. At the beginning of July 1954, the JTF-7 decreed:

> The routine welfare and care of all Marshall Island natives are continuing responsibilities of the High Commissioner, Trust Territory of the Pacific Islands, as agent for Department of the Interior. The Division of Biology and Medicine, AEC, will continue to monitor the physical condition of the native inhabitants of Rongelap and Utirik Atolls who were exposed to radioactive contamination as a result of the first shot of the CASTLE series. Parties of medical and radsafe personnel under the direction of the Divison of Biology and Medicine, AEC, will visit the natives and the atolls concerned periodically in order to observe the medical programs of the natives and to ascertain the earliest possible time for the return of the Rongelap natives to their homes. (E. McGinley to Manager, Santa Fe Operations Office, 6 July 1954.)

8. But it remained classified for months; the AEC did not want dose and distance information published in the literature, which would allow calculation of the yield from atmospheric tests.

9. Hardy et al. 1964.

10. "In our review of materials that are now becoming available, we found no evidence to support the claim that the exposures of the Marshallese, either initially or after resettlement, were motivated by research purposes. On the contrary, while there is ample evidence that research was done on the Marshallese, we find that most of it offered at least a plausible therapeutic rationale for the potential benefit of the subjects themselves. . . . We have seen no evidence . . . that convincingly demonstrates that research goals took priority over treatment in a way that would expose the populations to greater than minimal risk" (Faden 1996, 368, 372). The team's primary mission, Conard wrote, "was to treat the people. I don't think at any moment the motivation . . . was anything other than treatment of the effects of radiation." Though he did also write that "we [also] were trying to get as much information as we could into the medical literature. We knew that we were dealing with an area that was unexplored in human beings and we wanted to find out as much as we could about" the effects of exposure (Conard 1980).

11. Konrad Kotrady, unpublished.

12. Robert A. Conard to Edward E. Held, June 6, 1958.

13. Mike Malone, "Administration Medical Team Help Refused," *Pacific Daily News*, 28 March 1972). Conard had been handed a note: "TO A.E.C. MEDICAL TEAM: We are sorry the people on Rongelap were not able to cooperate with the A.E.C. medical team in the examinations this year. We were advised not to cooperate by our Congressman," Memo, Rongelap Council, March 16, 1972.

14. Thus the Marshallese politicians, too, were not immune to exploding coconuts. Though Lekoj's illness might still have been fatal, earlier examination might have detected the disease and extended his life. The interruption in the examinations had led to the one person who was most seriously ill being denied his best chance at treatment.

15. Knudson letter.

16. Resolution 28, Nitijela of the Marshall Islands, 17th Constitutional Regular Session, September 10, 1996.

17. Ibsen to [Frederik V.] Hegel, 21 June 1882.

References

Advisory Committee on Human Radiation Experiments (1996). *Final Report of the Advisory Committee on Human Radiation Experiments.* New York: Oxford University Press.

Benford, Robert D. (2002). "Controlling Narratives and Narratives as Control within Social Movements," in J. E. Davis, ed., *Stories of Change: Narratives in Social Movements.* Albany, N.Y.: State University of New York Press.

Conard, R. A., D. E. Paglia, P. R. Larson (1980). Review of Medical Findings in a Marshallese Population Twenty-six Years after Accidental Exposure to Radioactive Fallout. BNL 51261.

Cronkite, E. P., Bond, V. P., and C. L. Dunham, eds. (1956). *Some Effects of Ionizing Radiation on Human Beings.* Atomic Energy Commission TID 5358.

Crease, Robert (1999). *Making Physics: A Biography of Brookhaven National Laboratory, 1946–1972.* Chicago: University of Chicago Press.

————. "Conflicting Interpretations of Risk: The Case of Brookhaven's Spent Fuel Rods." *Technology: A Journal of Science Serving Legislative, Regulatory, and Judicial Systems* 6 (1999): 495–500.

Faden Ruth R. ed. (1996). *The Human Radiation Experiments: Final Report of the President's Advisory Committee.* New York: Oxford University Press.

Fairchild, Amy L., and Bayer, Ronald (1999). "Uses and Abuses of Tuskegee," *Science* 284:91–3.

Geertz, Clifford (2000). *Available Light: Anthropological Reflections on Philosophical Topics.* Princeton, N.J.: Princeton University Press.

Hacker, Barton C. (1994). *Elements of Controversy : The Atomic Energy Commission and Radiation Safety in Nuclear Weapons Testing 1947–1974.*

Hardy, Edward P., Jr., Joseph Rivera, and Robert A. Conard (1964). "Cesium-137 and Strontium-90 Retention Following an Acute Ingestion of Rongelap Food." Upton, New York: Brookhaven National Laboratory Technical Report #8657.

Hewlett, Richard G. and Holl, Jack M (1989). *Atoms for Peace and War 1953–1961.* Berkeley: University of California Press.

Kellerman, Jonathan (1995). *The Web.* New York: Bantam.

Kotrady, Konrad (unpublished). "The Brookhaven Medical Program to Detect Radiation Effects in the Marshallese People: A Comparison of the Peoples' vs the Program's Attitude."

Neumann, Klaus (2000). "Starting from Trash," in Borofsky, ed., *Exploring Pacific Pasts: An Invitation.* Honolulu: University of Hawaii Press.

Wesley-Smith, Terence (2000). "Historiography of the Pacific: The Case of The Cambridge History," in *Race and Class* 41(4): 101–17.

Classifying People: Science and Technology at Our Service

Essentially Empirical: The Roles of Biological and Legal Classification in Effectively Prohibiting Genetic Discrimination

Anita Silvers and Michael Ashley Stein

Introduction: Genomics and the Call for Justice

Science promises enormous public benefit, but simultaneously threatens significant social harm. Genomics is no exception to the enigmatic prospects of science. Genomics, the study and application of genetic information, can identify asymptomatic individuals who are at risk of becoming ill or of transmitting inheritable illnesses to their children. In principle, people who test positively for potentially disabling genes could take various kinds of prophylactic measures to slow or stop manifestations of disease. At the same time, however, predictive genetic testing could make unprecedented numbers of asymptomatic people vulnerable to the kind of workplace discrimination usually targeted at people with detectable disabilities.

Whether, on balance, the changes wrought by genomics prove of more help than harm will depend on how effectively we can temper the use of genetic information with justice. In testimony before the Congress, Francis Collins, Director of the U.S. National Human Genome Research Institute, called for policy that will dispense justice to individuals whom genetic tests reveal to have biological anomalies: "While genetic information and genetic technology hold great promise for improving human health, they also can be used in ways that are fundamentally unjust. Genetic information can be used as the basis for insidious discrimination."[1] Genetics research studies are already being deprived of subjects, he said, because prospective participants fear that their genetic profiles "will fall into the wrong hands and be used to deny a job or promotion."[2]

Collins asked, "Should employers be allowed to use genetic information about misspellings in genes, whether of a carrier or of an individual with misspellings that may confer increased risk of disease, to deny employment, even though the ability of the individual to perform the essential functions of the job is unaffected?"[3] Political leaders from Bill Clinton to George W. Bush joined Collins in answering "No!"[4] As Collins observes, "It is estimated that all of us carry dozens of glitches in our DNA." So Bush and Clinton, Collins and Venter,[5] and you and I, all seem to have a common interest in there being strong legal protection against genetic discrimination in the workplace.

With so much support for taking action, there surely was reason to expect that progress against genetic discrimination would come at least as quickly as progress toward knowledge of our genes. In the months after the announced mapping of the human genome, the actions of employers believed to be using genetic information to disadvantage workers[6] provoked considerable public outcry. There can be little doubt among lawmakers as to public sentiment. Nevertheless, the general population is no better protected from genetic discrimination today than in June 2000, when the mapping of the human genome was pronounced a success, or in July 2000, when Collins called the dangers of genetic discrimination to the attention of the Congress.

The failure to act effectively is not easily explained away as indecision or weakness of legislative will. Attempts to develop protective legislation encounter the prevailing presumption—deeply embedded in cultural practices and thought—that to depart from biological species-typicality is to deserve disadvantage.[7] As we shall see, prevailing legal conceptions enshrine this presumption. For instance, the U.S. Supreme Court has declared that laws may treat biological minorities differently without breaching equal protection as long as the "classification drawn by the statute be rationally related to a legitimate state interest."[8] It is to this standard of equal protection the Court continues to revert when a minority's biological atypicality incurs social disadvantage.[9]

In this chapter we explore presumptions about the classification of genetic minorities. They are an important source of resistance to developing legal protections against genetic discrimination in employment, the kind of discrimination Collins decried in his testimony. First, we identify a principle that constrains public remedies for deterring discrimination and consider arguments for its applicability to genetic discrimination. Second, we point out that the influence of this principle has been curtailed in order to ban discrimination against any citizen if the discriminatory conduct is based on race or sex. Third, we review current legislative approaches to prohibiting genetic discrimination and show that they have not been similarly unfettered. Contrasts between how we construct membership in a nondominant race or sex and membership in a genetic minority emerge from this discussion. Although biological differences are as central to drawing racial and sexual classifications as to genetic classification, the approach to classification is not the same in the former cases as in the latter one. Finally, we suggest how to reconceptualize the classification of people who are members of genetic minorities in order to provide more effective protective legislation for them.

At-Will Employment

Proponents of neoclassical labor models embrace the principle of at-will employment, and courts very often take it as a base-line in employment discrimination cases.[10] On this theory, public policy that restrains employers from hiring and firing employees at will invites significant efficiency losses. At-will employment permits employers to hire and fire on the basis of any properties of workers which attract or repel them.[11] Interfering with the selection process of fully informed employers is supposed to impose unwarranted costs on them (having less than optimal workforces), on optimally deserving workers (being overlooked in favor of less deserving candidates), and on the public collectively (suffering the consequences of flawed production).[12]

On this principle, in its categorical form, employers may fire or refuse to hire employees on the basis of racial, sexual, or genetic identities, regardless of whether they personally are well qualified for and productive in the job. If employing minority employees makes for a more productive workforce than not employing them, rational employers are supposed to abandon their exclusionary practice. Assuming employers to behave rationally, their retaining exclusionary practice is taken to warrant its virtue. On this view, the persistence of racial discrimination does not refute the claim that rational market forces correct perverse practices.[13] In the absence of substantiated links of racial characteristics to defective productivity, racial discrimination is accounted to be a matter of employers using race as a proxy for less easily perceived deficits in productivity, or of society's rewarding "tastes" or preferences for discrimination.[14] Racial discrimination is therefore characterized as a rational response to forces found in the market.

The principle of at-will employment must prevail, it is argued, in order to permit market forces to do their work. There is no countervailing rational or moral demand for employers to develop inclusive workforces. How does the principle of at-will employment play out in regard to genetic properties?

If efficiency is of primary concern, and if fully informing and unfettering employers results in their assembling optimally efficient workforces, it follows that they deserve to learn about the genetic profiles of prospective and present employees and to be free to act on this information. For instance, law professors Colin Diver and Jane Maslow Cohen propose that "armed with genetic test results and corresponding epidemiological data on the correlation between genotype and phenotype, employers may be able to improve the quality of the predictions they can make about the two determinants of job performance: intensity and quality of effort."[15] They suppose that "genetic information may someday provide a more reliable basis for measuring deficits in job-relevant skills that can be corrected by the design of training programs."[16] These authors deny genetic determinism,[17] but they fail to provide an alternative model that would support their claims about predicting the quality of individuals' job performance on the basis of inherited characteristics of their phenotypes.

Proponents of making law an instrument of economics are not alone in advocating that genetic information should play a powerful role in employment decisions. The influential philosophers who wrote *From Chance to Choice* believe that not efficiency, but justice, calls for assembling optimally productive workforces.[18] Equality of opportunity is meaningful only if individuals can develop and reap the rewards of their different skills and talents. Social opportunities emerge within a frame of cooperative schemes that require collective participation. How such schemes are structured affects who can participate, who is likely to succeed, and the quality of the products achievable.

Justice is denied, say the book's authors, if people with greater physical or cognitive potential are denied the challenges and compensations available in a more strenuous cooperative scheme and instead are relegated to a more inclusive, but less rewarding, social order. Justice owes people with greater capability the opportunity to participate in a workforce with an appropriately high threshhold of eligibility. These authors argue that inclusive cooperative schemes may unjustly impede species-typical individuals, who are in the majority, from fully developing their talents. Inclusive workplace arrangements, designed to protect the least able from failure, may reduce the rewards available to the most able. Information about people's genetic identities can serve justice, in this view, by facilitating the selection of groups of workers with similar abilities and prospects, and by identifying currently asymptomatic individuals whose inclusion in a workforce may eventually prove burdensome for their more capable colleagues.

In sum, the principle of at-will employment supports employers using genetic information to avoid employing individuals who currently are optimally productive but who may in future be compromised by genetic disease. If at-will employment is the prevailing principle, employers are free to assemble workforces by selecting for the genetic profiles they prefer. Further, the principle facilitates employers passing over genetically disadvantaged people in order to create an homogenous workplace that challenges and rewards genetically favored employees.

If the principle of at-will employment prevails without constraint, the benefit to a worker of acquiring genetic self-knowledge may be overwhelmed by disadvantage that is imposed by employers or others. The principle makes it permissible, and perhaps even obligatory, for employers to exclude individuals they suppose to be genetically flawed from the workforce. In these circumstances, it is rational, and possibly a prudential obligation as well, for individuals to refuse genetic testing, even if being tested might contribute to the individual's health or the health of others. Given the courts' deference to the principle of at-will employment, and absent a climate of confidence in the prospect of constraining genetic discrimination, there is reason to be pessimistic about the likelihood of realizing genomics' beneficial promise.

Constraining Discrimination

Over the past half-century, prohibitions against discrimination based on race or sex have gained legal power to abridge the principle of at-will employment. In general, employers may not privilege nor disadvantage current or prospective employees on the basis of race or sex, unless having or lacking a particular racial or sexual identity can be demonstrated to be essential to successful performance of the job.[19] Further, employers adopting exclusionary policies must show why the presumption that differences of race or sex are not also differences in competence to work may be abrogated in their case. For example, an employer cannot ban all women from employment as firemen, even if the majority of women don't qualify on strength and stamina performance tests. Further, tests that tend to exclude women must be tightly tied to the essential functions of the job. Thus, if removing victims from a dangerous site is important to a job, if victims may be removed by being dragged instead of being carried, and if women generally cannot carry victims but can drag them, applicants cannot be required to pass the "fireman's carry" test as a condition of qualifying as firemen. Instead, qualifying tests for appointment to this job must be revised to eliminate any unjustified disparately disadvantageous impact on women. Of course, a few jobs, such as sperm donor or egg donor, are tightly tied to being of a certain sex. Refusing even to consider the individual qualifications of biological females who apply to be sperm donors is not prohibited by the interdiction on sex discrimination.

Courts have recognized several compelling reasons for combating blanket workplace exclusions that target race or sex. There is, first, the social importance of affording meaningful and equitable employment opportunity to members of both sexes, and of all races, for neither sex nor race excuses individuals from obligations to be self-sufficient and to provide their dependents with support. Second, the long history of oppression of people of color and women cannot help but have shaped current practice so that their disadvantage persists today. Third, people irrationally dislike interacting as equals with individuals of a certain race or sex and have irrational attractions and aversions to individuals of a certain race or sex performing certain roles. Finally, race and sex often become proxies for other properties, especially ones that are undesirable for workers to possess. For instance, classification as female has been taken as a proxy for seductiveness and frailty. Similarly, having African forebears has been taken as a proxy for indolence and simple-mindedness. For these reasons, even theorists who place their faith in the unfettered market's compelling rationality have agreed to constrain employers' liberty to discriminate on the basis of race or sex.[20]

Next we should ask whether there are similar reasons for constraining employers from exclusionary practice based on genetic profile. It is, surely, as important for genetic minorities to gain self-sufficiency, and for their families to obtain adequate support, as for women and people of color to enjoy these achievements. Surely, as well, everyone will benefit if people with greater than typical risk of developing an illness are in a position to save money and otherwise prepare themselves with resources for their own care if taken ill.

There can be no doubt, further, that in the past the specter of inherited weakness or disease has exposed persons believed to be so marked to oppression. Most of us are aware of the atrocities carried out in the name of eugenics. These programs were conducted with the goal of enhancing the level of collective human performance by removing underachieving performers, for instance, through the involuntary sterilization or termination of putatively warped and burdensome kinds of people or the banning of immigration from parts of the world designated by the 1924 U.S. Immigration Restriction Act as having "biologically inferior" indigenous populations.[21] Sterilizing blind Californians and enjoining immigration from Eastern Europe because feeble-mindedness was supposed to be endemic among Jews are just two of the exclusionary policies that have been practiced in the United States within the memory of living citizens.[22] Beginning in the middle of the nineteenth century, citizens believed to have inheritable cognitive deficiencies, who had largely taken care of themselves, were expelled from workplace and community and segregated into custodial institutions.[23] As Supreme Court Justice Thurgood Marshall observed, "a regime of state-mandated segregation and degradation soon emerged that in its virulence and bigotry rivaled, and indeed paralleled, the worst excesses of Jim Crow."[24]

Law professors Diver and Cohen argue that, "given the enormous diversity of genetic characteristics and the hidden nature of most genotypes, one cannot plausibly assert that genetic aversion is an argument in the preference functions of most people."[25] Their conclusion appears to rest on a mistake about the standard of evidence for explaining how people's preferences are influenced. People may be aversive to interacting with members of a group with a particular genetic profile, and yet interact with individuals who belong to the group because they are ignorant of their profiles. Indeed, the authors themselves suppose that the current behavior of employers differs from how they would behave if they had such knowledge. Thus, the current level of aversive conduct based on genetic profiles is no evidence that the level would not naturally rise were people to become privy to more genetic information about one another.

There should be no doubt that people experience compelling aversions to certain individuals because they perceive the individuals' putative defects to be inherited and therefore inheritable, rather than accidently acquired. Inheritable anomalies are especially alarming because they appear to compromise the future. For example, when Justice Oliver Wendell Holmes referred specifically to mental retardation (or, more accurately, to a condition that was mistakenly diagnosed as such) in his famous remark in *Buck v. Bell* that "three generations of imbeciles are enough,"[26] he was reacting aversively to the prospect that the condition was inheritable. He endorsed sterilization precisely because it arrests the intergenerational transmission process that is the hallmark of genetic disease.

So far, we have seen that genetic discrimination resembles race and sex discrimination in respect to being fueled by irrational aversion, embedded in practices shaped by a long history of oppression, and antipathetic to the promotion of a population of self-sufficient, productive people. There is, however, yet another feature of race and sex discrimination, one on which genetic discrimination's resemblance to race and sex discrimination may be harder to make out. One of the most invidious aspects of the latter conduct is that membership in biological minorities, or nondominant biological groups, becomes a proxy for attributions of socially undesirable properties, in disregard of the presence or absence of empirical evidence linking these properties to the group.

Genetic discrimination unquestionably can involve the unsubstantiated linking of biological properties or conditions with undesirable social behavior. To illustrate, the sterilization of Carrie Buck, the unsuccessful plaintiff of *Buck v. Bell*, was carried out on the supposition that she had inherited her mother's and grandmother's mental retardation and because of the conviction that retardation is a proxy for immoral conduct. The convention that retarded people are immoral people prompted Justice Holmes to hold that states have the right to prevent retarded people from reproducing.

It may be argued that genomics itself offers an antidote to the use of genetic identities as proxies. For the error that permeates *Buck v. Bell* arises from the conviction that people inherit, rather than acquire, dispositions for illicit conduct. Eventually, genomics should be able to tell us which, if any, elements of human behavior are the result of biological inheritance. In principle, therefore, genomics could eliminate the unsubstantiated extrapolations that result in genotypes being made into proxies and thereby becoming instruments of discriminatory practice.

The Nature of Proxies in Legal Methodology

In practice, does such discovery of facts correct the erroneous proxy process? To answer, we need a better understanding of its nature and of how it can be overcome. Once, being a woman was a proxy for being incapable of executing traditional male roles. So it will be useful to see what makes a proxy, in order to see how proxies can be unmade.

In *Goesart v. Cleary*, in 1948, the Supreme Court upheld as obvious the constitutionality of a Michigan statute prohibiting the licensing of women as bartenders (except spouses or daughters of male bar owners).[27] Michigan was deemed reasonable in excluding women from this profession. Despite acknowledging that the preceding war years had brought "vast" social and legal changes to women's status, the Court held that to restrict a profession mainly to men by state action does not violate the Constitution because equal protection does not mean equal treatment for individuals whose situations are "different in fact or opinion."[28] No matter how adroit a woman might be at pouring drinks, or how competent at tallying sums, to the Court her situation could not help but be different from men's.

For the Justices, the mere thought of a female dispensing drinks evoked the image of a "sprightly and ribald" Shakespearean alewife[29] So, they believed, the mere presence of a female dispensing intoxicating beverages behind a bar could not help but raise "moral and social problems" which the state intended to prevent.[30] Only the "oversight" of a male with special interest in both the woman's welfare and the protection of bar room property—"assured through ownership of a bar by a barmaid's husband or father"—could be trusted to minimize "hazards" otherwise confronting an unprotected barmaid.[31] Because the line drawn by the Michigan legislature was not wholly lacking reason, the Court held that the disadvantage the statute imposed on most women did not call its constitutionality into question. Equal protection consideration could not reach this purpose, however "unchivalrous"[32] or exclusionary it might be, because the state had a rational interest both in protecting women from the limitations of their ability to defend themselves and in protecting the public from disruptions provoked by the mere presence of women in a potentially raucous uncontrolled environment.[33]

Undoubtedly, some women cannot impose order on the male patrons of a bar, and possibly this deficit is more frequently found in women than in men. Of note in this regard is the *Goesart* Court's decision to limit the opportunities of the class's members categorically because of the deficits of some of its individual members. In *Goesart*, the overt issue is whether the excluded class's members are capable of performing certain roles, but there is an underlying issue about whether the state has a rational interest in their not doing so.

Taking a class's biological difference as a proxy for its members having atypically limited ability defeats claims to equality. For, arguably, there is no value in providing two groups with similar opportunity if one of the groups has differences that keep members from successfully pursuing it. Historically, courts have addressed the constitutionality of limiting opportunity for classes delineated in terms of biological differences by considering two related questions: First, does the class members' biological difference relate to or signify some type of "reduced ability to cope with and function in the everyday world," so that class members generally need special protection rather than similar opportunity?[34] Second, does the class members' reduced ability to cope and function usually place the public in need of special protection?

Answering either of these questions appears to turn on facts about the conditions under which either the class members or the public need special protection, the frequency with which such conditions might occur, and the degree to which they can be averted. These questions all are empirical. Whether women can defend themselves and quell disruptions their presence may provoke unquestionably is decidable by observing what women do. Though there are women unlikely to evoke raucous reactions in men, and there also are women capable of quelling drunken disturbances, the *Goesart* Court believed women to be a general risk to themselves and the public if they tried to preside over barrooms. Although the Court should have treated this as an open question, with answers to be determined by establishing factual truths about the class, it instead took the question to be foreclosed by customary notions about the female sex.

Despite acknowledging "vast changes" that had occurred during the war years in women's social and legal position, the *Goesart* Court held that legislatures were not required "to reflect sociological insight, or shifting social standards," or "to keep abreast of the latest scientific standards" when drawing lines between the sexes.[35] The Court thus preempted answering empirical questions without examining the facts about women. This led the *Goesart* Court to consider women's situation sufficiently different to justify denying them opportunity for employment in bartending. It therefore held that it was rational to prevent them from doing so.

Fifty years later, very little, if anything, strikes us as obviously warranting employment discrimination on the basis of sex. Accordingly, the Court has stated that "the sex characteristic" does not reflect a woman's "ability to perform or contribute to society," and that sex-based distinctions characterizing relative capabilities are "outmoded notions."[36] Today, if the objective of a statute or practice is to exclude women in order to protect them, or to protect society from them, because they are supposed to be inferior, the objective itself is presumed illegitimate unless evidence that rises to a high standard of proof demonstrates otherwise. No longer a proxy for different and deficient capabilities, membership in the class of females is now, for purposes of equal protection, declared to be constitutionally neutral. During the half century that separates us from *Goesart*, the Court's characterization of the female classification in *Goesart* has come to seem "archaic and stereotypic" to the Court itself.[37]

Initially, we may be tempted to explain the transformation in the Court's account of the female classification solely in terms of empirical discovery, that is, as a process in which new evidence about women's capabilities came to light. This analysis misses an important point, however. In

Goesart, the Court explicitly declares that legislatures may construct legal classifications as they choose, especially if their characterizations of classes' members reflect convention, disregarding factual truths about the class. The reference to Shakespeare's ribald alewife that looms so prominently in the prevailing opinion in *Goesart* imposes, as a governing notion, a conventional cultural image of how women conduct themselves when serving drinks. In other words, it was facts about conventions about women, rather than facts about women, that the *Goesart* Court took as decisive in shaping their classification in the law.

Fifty years later, the Court's methodology in regard to classification by sex has changed. Now, empirical findings rather than conventional presumptions determine the attributes of the female classification. Once they adopted this basis, courts found no evidence that, in general, individuals classified as women had less ability to perform or contribute to society than men.

Since the nineteenth century, the Court has taken conventions, rather than facts, about groups as constitutive of the factual properties the law assigns to the group's members. The late nineteenth century decision in *Plessy v. Ferguson*, which affirmed racial segregation as meeting the equal protection standard, turned on whether an individual who could not be identified by others as "colored" could nevertheless be required to identify himself as such, and thereby be relegated to a segregated facility.[38] Plessy argued that his presence in a "white" railroad car could not be disruptive because he appeared to be white. However, Louisiana's classification of "white" and "colored" took the biological fact that he had an African forebear as a proxy for his color, although his skin was indistinguishable from individuals classified as white. The Supreme Court declared that states may construct racial classifications to reflect prevailing convention in their locales. Thus, one state might erect a "one drop of blood" standard for the "colored" classification, while another might designate that having a "colored" grandparent (rather than a more distant "colored" ancestor) constituted membership. Rather than being seen as natural biological groupings, racial classifications were declared to be established by statute, and race-defining properties assigned by man-made law rather than the laws of nature. Biological properties, such as being the grandchild of a particular individual, were made proxies for membership in these classifications, deceptively suggesting that biology endorsed convention.

As in the fifty years that followed *Goesart*, progress in the half century from *Plessy v. Ferguson* to *Brown v. Board of Education* involved a methodological change, a transformation from convention to empirical investigation as the basis for classifying races and determining their members' attributes.[39] For both racial minorities and women, there has been a methodological shift from reflecting custom to demanding empirical evidence. Once the evidence is permitted to be weighed, differences in race and sex no longer are identified with social contributions of different value. Indeed, denying that the performances of women and people of color are of lesser value has become a premise of constitutional law.

Not all classifications have become essentially empirical, however. In determining eligibility for the class protected against disability discrimination, courts tend to allow persuasive statutory definitions of disability—especially those drawn from welfare and rehabilitation law and designed to elicit generous allocations of resources—to trump neutral biological descriptions. Courts tend to take the assumptions about disabled people that typically inform such statutes—namely, that their performance is inadequate, and that they depend on receiving others' contributions rather than contributing themselves—as veridical of the class of disabled people.[40] Disabled people, thereby, are presumed to be a weak and needy class, regardless of whether many, a few, or none factually fit the description.

Unlike classifications of race and sex, attributions of impairments associated with disability continue to be accepted as proxies for social limitation. Genetic anomalies often are regarded as impairments. This raises a question about whether, for purposes of fashioning laws to bar genetic discrimination, genetic classifications resemble race and sex classifications less than they are like the disability classification.[41]

Approaches to Genetic Discrimination: Privacy Rights

In the main, two approaches to protection against genetic discrimination have been pursued. Initially, the appeal was to citizens' rights to privacy. More recently, antidiscrimination safeguards have been invoked. There is an important difference between the former and latter routes. Securing privacy proceeds on the assumption that everyone is equally vulnerable to intrusion, while prosecuting discrimination identifies and embraces those especially at risk.

Genetic information about an individual is discovered in several different ways. A remark about family history may reveal significant data. Data are accumulated in a medical setting, where informed consent for obtaining it is, in principle, necessary. In practice, however, patients often are asked to consent only to contributing a specimen for certain panels of tests. Or they are informed of the tests to be run without specifying what is learned from the tests. The physician may order the panel for one reason, which she discusses with the patient, but the entire set of test results becomes part of the patient's record.

Several difficulties about the control of genetic information arise. First, whose responsibility is it to identify or safeguard sensitive and easily portable genetic information? In many businesses, individuals who administer health care benefits or manage health and safety programs also have responsibility for some aspects of managing personnel. In these circumstances, is it feasible to expect employers to maintain a firewall between health care records that may reveal employees' genetic conditions and information used in personnel decisions?

Second, when someone waives a privacy right for one purpose, is it possible to keep the information protected from other purposes? A genetic anomaly which is correlated with one condition may, in future, be correlated with another, or anomalies may cluster, so that the presence of one suggests the presence of another. To illustrate, individuals who provided DNA to be tested for susceptibility to heart disease might, years later, find that their physicians have recommended suspension of their drivers' licenses because of new data that the gene correlated with heart disease also is associated with Alzheimer's Disease.

Third, genetic make-up is shared among close biological relatives, so test results for one person can yield information about another person. In such cases, do we defer to the individual who will benefit from disclosure or to the one who wishes to preserve privacy? The proprietary model on which an individual's medical information is her own private property ill fits genetic information.

As the real difficulties of protecting genetic information have become more evident, enthusiasm for basing protection on privacy rights has waned. Attention has turned to the antidiscrimination model already functional in federal disability discrimination law. Where the privacy model extends protection by sequestering information, the antidiscrimination model assumes that such attempts may be unsuccessful, and, consequently, regulates the uses to which genetic information may be put. In the United States, discussion has centered on whether the Equal Employment Opportunity Commission's application of the Americans with Disabilities Act (ADA) to genetic discrimination is adequate or whether it is necessary to develop separate protection for people with genetic anomalies.

Approaches to Genetic Discrimination: The ADA

Federal courts have required individuals who hope to be safeguarded by the ADA to prove that they have disabilities. Being disabled means having "(a) a physical or mental impairment that substantially limits one or more of the major life activities of such individual, (b) a record of such an impairment; or (c) being regarded as having such an impairment."[42] Human Genome Project Director Francis Collins has remarked that "it is estimated that all of us carry dozens of glitches in our DNA. . . . As a nation, we have stated unequivocally" in the ADA "that one's ability to do a job should be judged on just that—the ability to do the job."[43]

Initially, the ADA appears to have potential for protecting against genetic discrimination. The ADA does not specify application to genetic conditions. There are, however, several reasons for thinking it is applicable, at least to some extent.

First, the Congressional Record offers some evidence of legislative intent. For example, Congressman Major Owens stated that "[t]hese protections of the ADA will also benefit individuals who are identified through genetic tests as being carriers of a disease-associated gene. . . . The determination as to whether an individual is qualified . . . may not be based on speculation regarding the future."[44] In sum, congress members characterized genetic discrimination as exhibiting the myths, fears, and stereotypes that historically have excluded people perceived as biologically anomalous from fair equality of opportunity.

Second, the EEOC has offered guidance in regard to actions arising from genetic information relating to genetic disease or disabling conditions. The EEOC's position as explained by Commissioner Paul Miller is that

> [a] person is "regarded as" disabled within the meaning of the ADA, if a covered entity mistakenly believes an individual has a substantially limiting impairment, when in fact, the impairment is not so limiting. Under such a theory, coverage for individuals with a genetic predisposition would generally rely on demonstrating a mistaken belief concerning the major life activity of working.[45]

Third, the ADA clearly protects individuals with inherited impairments such as muscular dystrophy, retinitis pigmentosa, osteogenesis imperfecta, achondroplasia, and Williams syndrome. Regardless of the degree to which they are symptomatic, individuals with these genes have the inherited conditions. Some are compatible with a range of limitations. For instance, the skills of people with Williams vary.[46] Some are classified as mentally retarded while others attain college and graduate degrees. A state that proposed to sterilize all its citizens with Williams (as some states used to do)[47] very likely would be charged with disability discrimination under the ADA. In that event, it would be exceedingly odd if the ADA protected only people with Williams who are unable to finish high school, but that college graduates with Williams had no defense against state sterilization programs because the gene showed less expression in their cases.

Some of the genetic conditions referenced above—for instance, muscular dystrophy and retinitis pigmentosa—are progressive. Individuals may be asymptomatic when they test positive for these genes, yet face a future of substantial limitation. Are such individuals protected by the ADA while they are asymptomatic? Suppose an employer believes, mistakenly, that an individual who is blind cannot perform a particular job. It would be disquieting if employers were permitted, on the basis of genetic information, to exclude qualified individuals with the retinitis gene who could see perfectly well, but banned from excluding individuals who had already lost their sight because of their retinitis.

Clearly, Congress intended to protect citizens who are discriminated against here and now because other people fear the future effects of the disease for which they are at high risk. Nevertheless, there is a question about the propriety of doing so by calling these citizens "disabled." Presymptomatic people often reject the idea of being assigned to the disability classification, even though disability discrimination is practiced against presymptomatic people. Further, the group of presymptomatic people who are vulnerable to disability discrimination could expand enormously as predictive genetic testing becomes more widespread. In this regard, Chief Justice Rehnquist's dissent in *Bragdon,* the Supreme Court's first ADA case, is suggestive. Rehnquist warned that "[r]espondent's argument, taken to its logical extreme, would render every individual with a marker for some debilitating disease 'disabled' here and now because of some future effects."[48] Indeed, courts have been concerned to interpret the ADA so as to limit the number of people who fall under its protection.

Plaintiffs may proceed under the "regarded as" prong of the ADA. To establish a claim of being regarded as disabled under the ADA, they must demonstrate that their employer mistakenly believes they have a physical or mental impairment which limits a major life activity, when they in fact have no such limitation. Congress extended the ADA's definition of disability to this group of functionally nondisabled individuals in order to combat erroneous but widespread socialized assumptions about people with "disabilities," what the Supreme Court in *School Board of Nassau County v. Arline,* a 1987 Rehabilitation Act decision, termed the "perception of disability based on myth, fear, or stereotype."[49] In *Sutton v. United Airlines, Inc,* the Supreme Court acknowledged the goal of protecting individuals misperceived as having disabilities that was articulated in *Arline.*[50] Nevertheless, the *Sutton* Court held that a defendant would have to entertain stereotypical misbeliefs that the plaintiff is ecumenically disabled—that she cannot perform an entire range of jobs in addition to the one from which she claims she has been unjustly excluded.

Predictive genetic testing typically occurs before the individual's genetic condition becomes symptomatic and causes substantial limitations of major life activities. Thus, we may expect that individuals who seek remedies for protection against genetic discrimination through the ADA often will claim that they have been treated unfavorably because they are regarded as disabled rather than because they are disabled. For example, Terri Sergeant, an individual with a family history of Alpha-1-antitrypsin, an often fatal deterioration of the lungs, has filed under the "regarded as" prong.[51] Sergeant was nonsymptomatic but tested positive for the genetic disposition for this disease, which had killed her brother at age 37.[52] As a result of the test, her physician initiated preventative therapy, which deters the development of the disease and protects against lung infection.

Of course, there is no certainty that Sergeant would have become symptomatic without the preventative therapy. Nevertheless, preventative medical intervention reduces or eliminates her risk, and she remains able to perform activities like walking and breathing, major life activities that are severely compromised in symptomatic cases of Alpha-1-antitrypsin. Nevertheless, when her employers learned of her condition, they fired her. Legally, however, to qualify for disability discrimination protection, she bears the burden of proof to show that her employers, who observed her daily, regarded her as currently unable to perform these activities, although the medical information indicated she would be able to do so as long as she has access to medical intervention. The record of litigation under the "regarded as" clause is insufficiently clear to know whether she will succeed under the ADA. It is unclear whether Terri Sergeant and other asymptomatic claimants can establish that employers regarded them as broadly disabled.

What is clear is the circular nature of their dilemma. To illustrate, positive genetic testing permits Sergeant to take preventative measures against the substantial limitations of major life activities that would result from Alpha-1-antitrypsin in the absence of medical prevention. Yet precisely because these measure have had success, she may be stripped of protection against losing her job and her medical benefits. Ironically, people may have to forgo the medical benefits genetic information can bring if they are to be protected by the ADA from discrimination based on that information. This catch-22 situation is not addressed by either existing or proposed statutory provisions, neither those that address disability discrimination nor those aimed specifically at genetic discrimination.

Approaches to Genetic Discrimination: Specifically Targeted

A legislative approach with many supporters in Congress focuses on prohibiting employment discrimination on the basis of genetic information. [53] The proposed legislation addresses predictive genetic information, acquired from the analysis of human DNA, RNA, chromosomes, proteins, and certain metabolites in order to detect genotypes, mutations, or chromosomal anomalies, or from information about genetic test results of, or occurrences of genetic disease in, family members.[54] Employers may not use predictive genetic information or information about requests for genetic testing or counseling to fail to hire, discharge, discriminate in working conditions or compensation, or segregate or limit employees in disadvantageous ways. But the legislation does not address any other kind of information.

There are several reasons to think that antidiscrimination legislation specifically targeted to protect the genetically anomalous population faces the same problems as antidiscrimination legislation that protects the disabled population. First, there is the matter of workers' ability to carry out the essential functions of the job. Employers can claim to be protecting workers who are at higher than usual risk of workplace-induced illnesses or injuries by excluding them from jobs that may harm them. Are workers who are regarded as needing such protection because of genetically linked toxic sensitivities unable to execute functions essential to the job because proximity to necessary work is a personal hazard? If so, should employers be required to continue their employment?

Second, the legislation's protection does not extend to important kinds of information. The Sergeant case illustrates the problem. Sergeant's employer could have gained knowledge about her genetic condition from several sources, not all of which qualify as protected. Data pointing to Sergeant's condition included the history of her sibling's illness and death, medical appointments to treat chronic respiratory problems that Sergeant attributed to an allergy, positive genetic test results for Alpha-1-antitrypsin, and medical records and bills for preventative treatment. The proposed legislation prohibits Sergeant's employer from basing an employment decision on the first and third items in this list, but not on the second and fourth. However, all the information the employer needs to identify her genetic condition is manifested in the record of her prophylactic treatment. An internet search will quickly identify the conditions for which the treatment is prescribed. Granted, knowing her family history might also offer a clue, but the employee would be hard put to establish that this protected information was the sole or crucial factor in the employer's decision.

Thus, the ban on using predictive genetic information does not systematically protect against unfavorable personnel actions that are prompted by beliefs about employees' dispositions to genetic illness. One of the main benefits an individual obtains from predictive genetic information about herself is the foreknowledge to take preventative or mitigating measures. Information that the employee is taking such measures is not protected. An employer concerned to eliminate workers with genetic susceptibility to asbestosis or mesothelioma from contact with asbestos fibers could identify behaviors that frequently occur when individuals learn of their susceptibility: ceasing to smoke, meticulous use of masks, and so on. On this basis, the employer could take action which would not be prohibited because it is not based on "genetic information."

Whether an asymptomatic individual is presymptomatic often is not clear. For example, a person who finds herself under stress and forgetting things might describe these circumstances to a physician. Forgetting things is no strict indicator of Alzheimer's disease, as witness the young parents who lock up their cars on sweltering summer days, forgetting that their infants are inside. In the case we are considering, the physician, knowing that this patient has a family history of early onset Alzheimer's disease, orders genetic testing, which gives a positive result for a gene associated with Alzheimer's disease.

However, an examination of the patient's cognitive functioning, with attention to the cognitive deficits diagnostic of Alzheimer's, is inconclusive. Although no diagnosis of Alzheimer's can be made on the existing evidence, the physician starts the patient on *Aricept* as a prophylactic to delay cognitive impairment, just in case the patient's memory problems signal the development of Alzheimer's. In this case, an employer who regards the employee as likely to develop Alzheimer's could claim to have based personnel decisions on inferences made from the unprotected parts of the medical record (the patient's report of memory problems and the prescription for *Aricept*) but not from the protected parts (the genetic testing and family history). As the Sergeant case and this case both show, prescribing medication to ward off onset of disease in individuals whom genetic tests show to be at risk may be as revealing as the test results themselves.

Under genetic discrimination legislation, individuals claiming harm from genetic discrimination most likely will have to establish that the harm occurred prior to there being any symptom of their condition. Questions about whether an employer's decision was influenced by unprotected parts of the medical record, rather than by the results of genetic tests or family history, may preempt bringing cases to trial. Individuals who use genetic information to pursue preventative measures to benefit their health may, in doing so, lose their legal recourse against genetic discrimination. Thus, the purpose of genetic antidiscrimination law—namely, to free citizens to improve their health through applications of genomic knowledge—may not be realized.

Analysis of these cases reveals a general theoretical difficulty in the application of antidiscrimination law to genetic discrimination, whether provided by the ADA or by specialized legislation. This is the problem of determining who will be protected and who not when no bright line separates vulnerable from impervious, and deserving from undeserving, populations. The ADA and the proposed specialized legislation each bifurcates the population into protected and unprotected groups. The ADA covers people who are symptomatic or mistakenly regarded as symptomatic. Specialized genetic protection covers asymptomatic individuals who, in former times, could have escaped discrimination but who now can be identified through genetic information.

Ironically, neither the disability discrimination approach, nor the attempt to provide separate protection from genetic discrimination, shields people who take mitigating measures to escape dysfunction. Further, the lines drawn between protected and unprotected groups do not reflect the difference between people who can and can't function successfully. Existing approaches to both disability discrimination and genetic discrimination thus fail in large part to reduce the costs of excluding otherwise productive citizens from equal opportunity if these citizens act to mitigate the effects of their biological anomalies.

Civil Rights in the Age of Genomics

The ADA has been read as extending civil rights protection to individuals whose physical or mental impairments substantially limit their participation in major life activities, or who are so regarded, but as giving no protection to individuals who can adapt to or mitigate their impairments sufficiently to fully execute such activities. On the other hand, proposed specialized legislation that targets genetic discrimination protects individuals until they take prophylactic measures against their genetic disposition. Here the protected population is asymptomatic and is almost a reverse mirror image of the symptomatic population protected by the ADA. Once again, however, individuals who take mitigating measures are unprotected. Both the ADA and specialized genetic discrimination law bifurcate the population into protected and unprotected groups, while leaving the large group of individuals who take steps to put off potential genetic disease unprotected from discrimination.

In contrast, no matter what their race and sex, regardless of whether they are identified with a dominant or an historically oppressed group, all citizens may, in principle, seek recourse through the law if they are harmed by race or sex discrimination. What would be required to take a similar approach and to extend genetic discrimination protection to the general population? Borrowing from the Civil Rights Act of 1964 (Title VII)—the central protection against race and sex discrimination—discrimination toward individuals on the basis of their genetic identity would be proscribed. Such a proscription would tailor genetic antidiscrimination protection to those instances when individuals have had their opportunities unfairly reduced because of stereotypic beliefs about the significance of the individuals' genetic identities.

Case law applying proscriptions against discrimination on the basis of race and sex now proceeds from the initial presumption that the prevalent characteristic of all protected individuals is their competence to perform, with a subcategory of individuals within the classification who will be unable to do so. This initial presumption will either be borne out or disproved by empirical evidence when particular actions are challenged. In line with our current treatment of racial minorities and women, the burden of proof in genetic discrimination cases should shift from requiring individuals who are anomalous to demonstrate that they can be competent and productive despite being anomalous to requiring whoever would exclude them from productive opportunity because of their anomalies to demonstrate that they cannot be.

Suppose, for example, that an employer refuses to hire an applicant because of her higher than species-typical risk of developing a genetic disease. An applicant whose genetic condition becomes so symptomatic in a year as to preclude her productivity is no different, the employer argues, from an applicant who announces she plans to quit in a year. Here the burden would fall to the employer to prove that the time horizons of employment in the two cases are of the same kind. But genomics itself precludes such proof, for knowledge that an individual has a gene associated with a disease is not knowledge of when disease symptoms will appear, even in the rare cases where it is knowledge that disease symptoms will appear because the gene in question has perfect (100 percent) penetrance.[55]

For purposes of the law, under our proposal, the population of the legal classification of genetically anomalous people would be characterized not in terms of cultural convention but, instead, through empirical study of the relevant biological groups. Through this methodological transformation, we would cease to use genetic anomalies as proxies for performance limitations. Of course, the science that has given us predictive genetic testing also has put us under threat, for genetic testing reveals properties that can be made into proxies, which are the instruments of discrimination. But the same science also offers the means for progressing beyond proxies to the substantiated probabilities associated with different genetic identities.

Just as women as a class now are thought able to protect themselves physically and to quell rowdy males, although some cannot do so, so people with higher than typical risk of genetic disease as a class should be presumed to remain viable employees, although some will not be so. The presumption should be that members of the class of genetically anomalous people will remain competent and productive, although a subclass will not be so, rather than that class membership means future deficiency, even though a subclass may escape this fate. Constructing the class of genetically anomalous people this way appropriately acknowledges that genomic knowledge supports judgments that are probable at best. Epistemological sophistication about assessing and ascribing risks on the basis of genetic identity will help in characterizing genetic classes for legal purposes.

This approach recognizes that in most cases genes associated with genetic diseases have less than 100 percent penetrance and also that many genetic diseases are multivariant, meaning that several factors must combine to induce the onset of symptoms. Individuals who are at higher than species-typical risk are nevertheless very often unlikely to become symptomatic. Further, even individuals who are symptomatic may maintain their competence and productivity, especially if mitigating measures for their disease can be found.[56]

There is at least one other feature this model requires. The standard of proof for excluding individuals on the basis of their genetic identities must present a reasonably high bar. Defending the exclusion of individuals on the basis of their genetic identities must be far more difficult to establish than a mere showing that their propensity to a genetic disease is more than species-typical. It will take very careful assessment of the epistemological characteristics of genetic information to tailor the appropriate level of proof required to exclude an individual. The requisite standard of proof must serve the liberty and opportunity interests of individuals and also satisfy collective social interests. The latter interests include both the reasonable desire of citizens to be self-supporting and the reasonable desire of employers to maintain productive enterprises.

Conclusion: Genetic Identity and the Route to Justice

We have argued for creating genetic discrimination protection "on the basis of genetic identity" for everyone, rather than just for "qualified individuals" who are symptomatic, or else who have positive results of predictive genetic testing but are not symptomatic. Everyone has a racial and a sexual identity, and a genetic identity as well. Further, everyone is genetically anomalous in some way. Everyone exhibits some differences from genetic species-typicality because the idea of species-typicality is as much an idealized construction as the idea of the "average man." Given these considerations, equality-based protection against genetic discrimination, with a scope similar to that for race and sex discrimination, is needed for everyone.

What science will discover about the problems attendant on each individual's genetic configuration, and which genetic configurations any employer may read as being proxies for unsuitability, is, at present, a lottery. Yet, medical research learns more and more every day about using genetic information beneficially to prevent or delay the onset of genetic conditions that may be disadvantageous. The population of the group that can take such mitigating measures is growing fast. Excluding these individuals from social opportunities cannot help but be enormously costly to society and to themselves, and, as well, to our faith that science can improve our lives. To save genomics, the major scientific achievement of our era, from concluding in such lamentable outcomes, we have proposed an approach to genetic discrimination that, because it protects everyone alike rather than designating a narrow protected class, would protect the people who have the most both to lose and to gain from genomics.

Notes

1. See Hearing Before the Senate Committee on Health, Education, Labor and Pensions, S. Hrg. 106–647 (July 20, 2000) (prepared testimony of Francis S. Collins).

2. Ibid.

3. Ibid.

4. For example, while praising the mapping announcement as having discovered the "book of life," President Clinton cautioned that "we must guarantee that genetic information cannot be used to stigmatize or discriminate against any individual or group." "Reading the Book of Life: White House Remarks on Decoding of Genome," *New York Times*, June 27, 2000, F8.

5. See "Lawmakers Renew Push for Gene Discrimination Bill," *Reuters Health*, June 7, 2001, available online at <http://www.reutershealth.com/archive/2001/06/07/elione/links/20010607elin037.html>.

6. Civil Action No. C01–4013 (N.D. IA.) (filed 2/9/01) (hereinafter Burlington Northern Complaint) (on file with authors).

7. Anita Silvers and Michael Ashley Stein, "Protecting Genomics' Promise An Equality Paradigm for Preventing Discrimination Based on Genetic Identity," *Vanderbilt Law Review*, (forthcoming).

8. 473 U.S. 432 (1985).

9. Anita Silvers and Michael Ashley Stein, "Disability, Equal Protection, and the Supreme Court: Standing at the Crossroads of Progressive and Retrogressive Logic in Constitutional Classification," *Michigan Journal of Law Reform*, v. 35, n. 1 & 2, Fall 2001/Winter 2002, 81–136.

10. See Colin S. Diver and Jane Maslow Cohen, "Genophobia: What is Wrong with Genetic Discrimination?" *University of Pennsylvania Law Review*, 149, (5) (2001): 1439–82, 1445.

11. Ibid. The legal doctrine of "at-will" in American law is derived from the "master-servant" provisions of English law. Absent a contract, both employer and employee have the right to terminate the employment relation at any time, for any reason. U.S. law imposes some restrictions on this right, notably by prohibiting employers from discharging whistleblowers, and from acting out of bias against a worker's race, sex, religion, or national origion.

12. Ibid.

13. Ibid.

14. Ibid.

15. *Op. cit.*, 1461.

16. *Op. cit.*, 1462.

17. *Op. cit.*, 1448, 1461–4.

18. Allen Buchanan, Dan Brock, Norman Daniels, and Daniel Wikler, *From Chance to Choice: Genetics and Justice* (Cambridge, U.K.: Cambridge University Press, 2000) 292.

19. Silvers, Anita, and Michael Ashley Stein. "From Plessy (1896) and Goesart (1948) to Cleburne (1985) and Garrett (2001): A Chill Wind From the Past Blows Equal Protection Away." In Linda Hamilton Krieger, ed., *Backlash Against ADA: Reinterpreting Disability Rights*. (Ann Arbor: University of Michigan Press, forthcoming).

20. Diver and Cohen, op. cit.

21. Troy Duster, *Backdoor to Eugenics* (New York: Routledge, 1990), 13.

22. Ibid.

23. Michael Burleigh, *Death and Deliverance: "Euthanasia" in Germany c. 1900–1945* (Cambridge [England] and New York: Cambridge University Press, 1994)

Marvin Miller, *Terminating the "Socially Inadequate": The American Eugenicists and the German Race Hygienists, California to Cold Spring Harbor, Long Island to Germany* (Commack, N.Y.: Malamud-Rose, 1996); Robert Proctor, *Racial Hygiene: Medicine under the Nazis* (Cambridge, Mass.: Harvard University Press, 1988); Dorothy Wertz, "Eugenics:1883–1970," *The Gene Letter* 3, 2, (February 1999a). <//www.geneletter.org/0299/Eugenics1883–1970.htm>.

24. *City of Cleburne v. Cleburne Living Center, Inc.* 473 U.S. 432 (1985) at 462 (Marshall, J., concurring in part and dissenting in part).

25. *Op. cit.*

26. *Buck v. Bell,* 274 U.S. 200 (1927), 207.

27. 335 U.S. 464 (1948).

28. *Id.* at 466 (quoting Tigner v. Texas, 310 U.S. 141, 147 (1940)).

29. *Id.* at 465.

30. *Id.* at 466

31. *Id.*

32. *Id.*

33. Presumably, the legislature's motive was to give returning male war veterans a monopoly on the occupation of bartending.

34. *City of Cleburne v. Cleburne Living Center, Inc.,* 473 U.S. 432, 442 (1985).

35. *Goesart v. Cleary,* 335 U.S. 464, 466 (1948).

36. *Mississippi University for Women v. Hogan,* 458 U.S. 718, 725 (1985) and *Cleburne,* 473 U.S. at 441.

37. *Id.*

38. 163 U.S. 537, 552 (1896).

39. See Silvers and Stein, "From Plessy (1896) and Goesart (1948)" *op.cit.*

40. See Silvers and Stein, "Disability, Equal Protection, and the Supreme Court" op. cit.

41. We argue that the disability classification also should be treated like the legal classifications of race and sex in Silvers and Stein, "From Plessy (1896) and Goesart (1948)." However, the grounds we give differ from our arguments for treating genetic identity analogously to race and sex identity.

42. 42 U.S.C. Sec §12,102 (2)(A)-(C).

43. See "Lawmakers Renew Push for Gene Discrimination Bill."

44. See 136 Cong. Rec. H4623 (July 12, 1990).

45. Miller, "Is There a Pink Slip in my Genes? Genetic Discrimination in the Workplace," *Journal of Health Care Law and Policy* 225 (2000). Although the article was written in his personal capacity, his view of the agency's position has also been reiterated in statements made in his authorized capacity. For example, Miller has stated that the EEOC "will continue to respond aggressively to any evidence that employers" misuse genetic information. See *Report Letter, EEOC Compliance Manual Report* No. 157 (April 27, 2001). Available online at <http://www.hr.cch.com/primesrc/bin/highwire.dll>. See also Prepared Statement of Paul Steven Miller Commissioner of U.S. Equal Employment Opportunity Commission Before the Senate Committee on Health, Education, Labor and Pensions (July 20, 2000) (Federal News Service).

46. See generally H.M. Lenhoff, et al., "Williams Syndrome and the Brain," *Scientific American* 277(68) (1997).

47. A comprehensive treatment of this topic is provided in Robert L. and Marcia Burgdorf, "The Wicked Witch is Almost Dead: Buck v. Bell and the Sterilization of Handicapped Persons," 50 *Temple Law Quarterly* 995 (1977).

48. *Bragdon v. Abbot,* 524 U.S. 624 (1998).

49. 480 U.S. 273 (1987). For a discussion of how courts should address this kind of error, see generally Michelle Travis, "Perceived Disabilities, Social Cognition, and 'Innocent Mistakes'" *Vanderbilt Law Review* 55, (2) (March 2002): 481–579.

50. 527 U.S. 471 (1999).

51. See Gene Johnson, *Update on Terri Sergeant's Genetic Discrimination Case,* available online at <http://www.alpha1.org/newsmakers/index.htm.>; *Genetic Discrimination is a Real Problem, With Real Victims,* available online at <http://www.nationalpartnership.org/healthcare/genetic/stories.htm>. Sergeant's story was first covered by the magazine *Scientific American* following her testimony before the Senate Health, Education, Labor and Pension Committee. *Id.*

52. Unless otherwise indicated, what follows is drawn from Johnson, *Update on Terri Sergeant's Genetic Discrimination Case.*

53. *See* Randall Mikkelsen, "Bush Proposes Ban on 'Genetic Discrimination,'" Reuters, (June 23, 2001), available online at <http://dailynews.yahoo.com/h/nm/20010623/ts/bush_genetics_dc_1.html>. The article's title and text are misleading. The president voiced support for an existing Bill, H.R. 602, introduced by Representative Louise Slaughter, rather than proposing legislation. A more accurate appraisal is David E. Sanger, "Bush Supports Federal Law Putting Limits on DNA Tests," *New York Times,* June 23, 2001, A10.

54. Ibid.

55. Very few disease related genes have 100 percent penetrance, meaning that disease always occurs when they are present. Huntington's Disease and Tay-Sachs Disease are examples of genetic conditions with 100 percent penetrance.

56. Lori B. Andrews, et al., eds., *Assessing Genetic Risks: Implications for Health and Social Policy* Washington, D.C.: National Academy Press, 1994.

Rethinking Normalcy, Normalization, and Cognitive Disability

Licia Carlson

The emergence of disability studies has raised many questions regarding the status, legitimacy, and effects of the medical model of disability. According to this model, disability is understood as the presence of certain physical or cognitive impairments, located in the individual, that are considered objectively "abnormal" and "undesirable." Along with the critique of an approach to disability that ignores the socially and historically determined nature of both the definitions and experiences of disability, the concept of "normalcy" has been called into question. A variety of what I shall call "normalcy critiques" have been formulated with respect to disability, and in most of these, medical science is implicated either explicitly or implicitly. A few examples from disability theorists include references to the "hegemony of the normal,"[1] the "disciplines of normality,"[2] "programmatic normalization,"[3] and the "tyranny of normalisation."[4] There are numerous ways in which the very concept of the normal can be critiqued: the ontological status of the categories "normal" and "abnormal" can be questioned; the binary nature of this concept can be challenged; and the practices associated with "normalization" of both disabled individuals and their environments have been called into question.

It is in the context of these normalcy critiques that I will consider a particular example of disability: mental retardation. Rethinking the concept of normalcy and the goals of normalization as applied to the case of cognitive disability raises a number of questions. First, what is problematic about the concepts of normalcy and normalization as they apply to persons with disabilities, and can this critique be extended beyond the medical model of disability? Second, can a critique of normalcy and normalization be applied univocally to the category of "disability" as a whole, or does the case of cognitive disability pose particular problems? In addressing the problem of specificity, I will discuss certain features of the category of mental retardation that suggest that the critique of normalcy must take into account the particular nature of the disabilities in question. Finally, in light of these reflections I will ask whether the concept of flourishing as presented in Alisdair MacIntyre's book *Dependent Rational Animals* (1999) may be preferable to the language of normalization and consider whether certain aspects of the normalcy critique can be applied to his analysis of disability.

The Normal/Abnormal Division: Ontological Questions

While critiques of normalcy have emerged in the field of disability studies, I am surprised that few disability theorists acknowledge the work of philosopher and historian of science Georges Canguilhem, in many ways a forerunner to the social model of disability. In his book *The Normal and the Pathological* (1991), Canguilhem considers whether claims regarding "normal" or "pathological" states in human beings can be "objective." He concludes, "There is no objective pathology. Structures or behaviors can be objectively described but they cannot be called "pathological" on the strength of some purely objective criterion."[5] Canguilhem goes on to argue that the category of anomaly is simply one of variation or diversity, and does not imply the designation "pathological: "In short, not all anomalies are pathological but only the existence of pathological anomalies has given rise to a special science of anomalies which, because it is science, normally tends to rid the definition of anomaly of every implication of a normative idea."[6] Canguilhem explains how, through the process of scientific inquiry and the production of scientific knowledge, naturally occurring anomalies/differences are deemed "abnormal" or "pathological." Thus, the line between "normal/abnormal" is not one found "in nature"; rather, Canguilhem argues that the concept of the normal is defined in part by virtue of the individual's relationship to the environment, and through a dynamic process:

> The normal is not a static or peaceful, but a dynamic and polemical concept. . . . To set a norm, to normalize, is to impose a requirement on an existence. . . . The reason for the polemical final purpose and usage of the concept of norm must be sought . . . in the essence of the normal-abnormal relationship. It is not a question of a relationship of contradiction and externality, but one of inversion and polarity. The norm, by devaluing everything that the reference to it prohibits from being considered normal, creates on its own the possibility of an inversion of terms. A norm offers itself as a possible mode of unifying diversity,

resolving a difference, settling a disagreement. But to offer oneself is not
to impose oneself. Unlike a law of nature, a norm does not necessitate its
effect. (Canguilhem, 240)

Canguilhem's work has profound implications for the medical model of
disability. By dislodging the notion that the abnormal/normal binary exists
objectively in nature, and by unmasking the process by which science covers its normative tracks, his work lends credence to the critique of the medical model of disability, a model that views disabling conditions as "objectively abnormal." Moreover, insofar as the normal is a dynamic concept,
Canguilhem suggests that it is in fact contingent and mutable, a fact evi-
denced in the complex history of changing definitions of "disability." Finally, because the norm does not in fact function as a "law of nature" that
necessitates an effect, an "inversion of terms" is made possible by virtue of
the normative nature of the polemical relationship between normal and
abnormal, thus leaving significant room for revision and resistance.

Canguilhem's claims are echoed in the work of philosopher of science
and disability theorist Ron Amundson, who discusses what he calls the "fal-
lacy of functional determinism," the belief in the objective biological reality
of the categories "normal" and "abnormal" function."[7] Amundson's work is
important because it challenges the ontological status of "normal" and "abnormal" kinds, and makes the further claim that this "myth of normality of
function" has served to oppress persons with disabilities. He writes, "The
reality of individual biological traits like blindness and paraplegia does not
entail the reality of the contrasting *kinds* normal versus abnormal function,
any more that the reality of diversity in skin colors and hair textures entails
the reality of the kinds called races."[8] Amundson draws upon biological sciences to illustrate the ways in which the traditional notion of some objective definition of "normal function" is challenged by examples of the adaptive powers of humans and non-humans to achieve functionality despite
"abnormalities." Amundson's critique calls into question the relationship
between definitions of "species typical function" and designations of particular kinds of individuals as abnormal. Furthermore, his examples suggest that the philosophical and scientific justifications for treating the concept of "normal function" as if it were an objective scientific fact found "in
natures" are problematic. He goes on to trace the implications this erroneous assumption has had for philosophical discussions of health care policy
and the treatment of persons with disabilities.[9]

Canguilhem's and Amundson's works call into question a purely medical model of disability that relies upon the reification of the normal/
pathological binary. Yet questions remain. First, while the myth of normal-

ity of function might be dispelled by pointing to the fact that the concept of the normal is a statistical one, thereby denying it any "objective" scientific status, this does not suggest that the concept of the norm and the normal/abnormal binary will cease to have meaning in both scientific and nonscientific contexts. Thus, I would argue that it is important to go beyond an examination of the ways that the medical model of disability reifies the categories of normal/abnormal and recognize that the force of this binary may persist in spite of such science, not simply because of it.[10] The importance of extending critiques of normalcy beyond the medical model becomes particularly evident when one considers cognitive disabilities. To illustrate this point, I will discuss three features of mental retardation as a classification that are relevant to arguments concerning normalcy and disability.

Mental Retardation as a Classification

James Trent, in his book *Inventing the Feeble Mind* (1994), argues that insofar as mental retardation is a problem, it is a problem of the mentally accelerated. His comprehensive history suggests that the perpetuation of the view that the mentally retarded are "abnormal" has largely been a function of the desire of nondisabled persons to draw and maintain this sharp dichotomy. Yet the complex nature of the classification of mental retardation itself raises interesting questions regarding the status of this dichotomy. In discussing three features of this classification, I hope to illustrate why specificity is required in developing a critique of normalcy as it applies explicitly to *cognitive* disability.

Amundson challenges the reality of normality and abnormal function as objective *kinds*, and thus the question of whether mental retardation exists as an objectively *natural kind* could be our point of departure. However, I would like to take a different path, in part because I have found the language of social construction unhelpful in discussing a category as complex as mental retardation. In one sense, there is no question that there are individuals who are labeled "mentally retarded" and who have significant cognitive impairments. However, in the spirit of Canguilhem's work, were we to ask whether the category "mental retardation" points directly to some objectively pathology found in nature, so to speak, the problem becomes more complex than simply asking whether mental retardation is "real" or not.[11]

The category mental retardation emerged in the United States around the mid-nineteenth century, when "idiocy" or "feeblemindedness" (the general heading for the category) was differentiated from mental illness through a series of political and scientific reforms. New schools opened for the "feebleminded," and the term went through a series of shifts, from feeblemindedness (which comprised the subcategories of idiot, imbecile, moral imbecile, and moron), to mental defectives, to "mentally retarded." If one examines the emergence, development, and perpetuation of this classification (it is still used today in a variety of settings), a strictly medical model for explaining mental retardation is both inadequate and reductionist. Three specific features of this category bear this out: its heterogeneity, its instability, and its ability to generate prototypes.

First, the category is both internally and externally heterogeneous. Internally, mental retardation has historically been, and continues to be, divided into multiple subcategories that are drawn along a variety of conceptual lines: etiology (e.g., genetic, environmental, social); severity (according to physical, genetic, and behavioral characteristics); and moral character (e.g., suggestibility, deviant or criminal behavior). It is externally heterogeneous insofar as it has been defined by multiple disciplines. It has historically been an object of knowledge in psychiatric, psychological, medical, educational, legal, and (less often) philosophical contexts. In fact, one of the most persistent features of this classification is the lack of consensus about the nature of mental retardation and the continual criticism of existing definitions.[12] Because of its location in multiple disciplines, then, to focus on the ways in which medical discourse about mental retardation reifies the normal/abnormal binary would reveal only a small part of the picture.[13]

Second, mental retardation can be discussed in terms of its stability. Unlike some other disabilities, the locus of the impairment has never been clear or undisputed. If we consider blindness as an example, vision is the locus of the impairment, and while definitions (legal and medical) of blindness have not remained stable, there is no question that blindness as a disability corresponds to some (albeit disputed or changing) definition of visual impairment. However, in the case of mental retardation, the definition of the impairment itself has changed and continues to be open to interpretation. At various points in history, mental retardation has been defined as a lack of intelligence, lower intelligence, negative will, suggestibility, inability to be educated, or moral defect. A recent genetic development locates the impairment at the chromosomal level (e.g., Fragile X retardation and Down's syndrome). In this respect I would argue that mental retardation as a category is unstable insofar as the very object of knowledge (that which is defined as a departure from a norm) continues to shift.

If one were to take Amundson's definition of the "myth of normality of function" as a starting point in addressing mental retardation as a disability, the question "Normal *what?*" is not as easily answered as in cases of blindness and paraplegia, for instance.

Mental retardation is an unstable classification by virtue of its heterogeneity. The final feature of this category—perhaps the most interesting one—arises from its instability: its ability to generate prototype effects. George Lakoff argues that the way we formulate categories depends upon features of our human cognition and experience and that our categories are often asymmetrical. He explains that "prototype effects" occur when one member of a category is judged more representative of a category than others.[14] The history of mental retardation offers many examples of this. At certain moments, the focus on a certain subgroup or definition has allowed a particular portrait of mental retardation to dominate. Specifically, the "mild" and "severe" cases have alternated in their prominence and in representing the category as a whole, and have functioned as prototypes both historically and in philosophical discourse.[15]

What is fascinating about mental retardation as a classification is its persistence. It is precisely because of, not in spite of, its heterogeneity, instability, and the ability to generate prototype effects, that it has been able to survive for so long. These features also point to the complexity of this category and raise serious doubts regarding any facile or general appeal to the broad category "cognitively disabled" or "mentally retarded." Moreover, the intricacies of this classification and its history suggest that a critique of normalcy as it applies to the category of "disability" generally is insufficient to address the problematic status of the normal/abnormal binary with respect to the definitions, perceptions, and treatment of persons labeled mentally retarded.

Normalizing Practices: Liberatory or Oppressive?

Thus far, I have been discussing possible critiques of normalcy at the conceptual level. However, much of the force of the disability critique of the normal/abnormal binary rests in the translation of this dichotomy into practice. In addition to the philosophical errors in perpetuating a myth of normality of function or in the reification of the "normal" and "abnormal" as absolute, objective kinds, disability theorists have critiqued the ways in which the medical model of disability has promoted and justified "normalizing practices." Critiques of normalization have focused on practices and strategies ranging from actions, social programs, and research devoted to restoring individuals to normal function, living, activities, or environments. Anita Silvers makes the connection between the concept of normalcy and strategies of normalization explicit when she writes:

> Normalizing equalizes opportunity only to the extent to which people can be maintained in or restored to the image of the dominant group. But we should recognize that the dominant group's fashions of functioning are not the product of any biological mandate or evolutionary triumph, nor are they naturally endowed to be optimally effective and efficient. Rather, members of this group impose on others a social or communal situation that best suits themselves, regardless of whether it is the most productive option for everyone. This being the case, we can see that the main ingredient of being (perceived as) normal lies in being in social situations that suit one—that is, in a social environment arranged for and accustomed to people like oneself. Thus, while in favorable situations normalizing can equalize individuals who are different by changing them, programmatic normalization—the equalizing strategy promoted by the medical model of disability—lends itself to oppression because it validates and further imposes the dominant social group's preferences and biases. (Silvers, 73–4)

What, then, are the implications of a disability critique of normalizing practices for mental retardation?

Historically, normalization has been a strategy explicitly tied to persons with cognitive disabilities. The program to integrate persons into the community and to teach them "normal" life skills and modes of being can be traced through the second half of the twentieth century. Wolf Wolfensberger, one of the central figures in the normalization movement, describes it as follows: "The most explicit and highest goal of normalization must be the creation, support, and defense of valued social roles for people who are at risk of social devaluation."[16] In many ways, the goals of normalization can be seen as a positive development in the history of mental retardation, as a means by which persons with cognitive disabilities might be less marginalized and become socially valued. However, an important distinction must be made between normalizing persons and normalizing environments.

The critique of the medical model of disability has raised challenges to both. Insofar as the approach to disabling conditions has been to "fix" or "cure" them, many disability theorists are suspicious of focusing normalization on individual bodies. The question of normalizing persons becomes interesting in the context of cognitive disability because the locus is not necessarily medical but pedagogical; the site of reform is not physical but behavioral or cognitive. While normalization in the pedagogical setting is far too complex an issue to be addressed here, it is important to note the vigorous debate regarding mainstreaming and inclusion of both mildly and severely disabled individuals in schools. Some critics have argued that, despite efforts to the contrary, these "normalizing strategies" have not served children with disabilities well.[17]

In addition to concerns regarding normalizing strategies in education, some disability theorists are also wary of other attempts to normalize individual behaviors and environments for the cognitively disabled. Eva Kittay writes, "We have to think carefully about what norms of 'normal' we want to subscribe to. Where the norm is 'independent living' rather than a caring environment wherein individuals thrive, we have to be cautious."[18] Richard Jenkins echoes this concern in his article "Toward a Social Model of (In)Competence," when he writes, "The politics of equality have inspired the modern philosophy of care known as normalisation. This promotes independent living for people with intellectual or other disabilities, in ordinary community settings wherever possible, and their participation—once again as far as possible—in culturally normative behaviors and activities. Ideologies of earthly equality do not, however, necessarily solve the problems faced by people with intellectual disabilities. They may, in fact, be reshaped into a new kind of oppression: the tyranny of normalisation may be as powerful as the opportunities offered by egalitarianism."[19] While there are reasons to question the motives and limits of medical or scientific normalizing strategies, the external heterogeneity and instability of mental retardation as a classification also point to the need for further examination of normalizing strategies as they apply to particular groups of individuals with cognitive disabilities in non-medical contexts.[20]

Dependent Rational Animals and the Possibility of Flourishing

My discussion of the discourse of normalcy and normalization has been quite broad, and given the external heterogeneity of the category of mental retardation, a comprehensive analysis of how normalcy critiques might apply to cognitive disability requires going beyond the boundaries of philosophy. However, I would like to dwell in the philosophical world a bit longer and conclude with a brief consideration of MacIntyre's *Dependent Rational Animals*. His articulation of the relationship between dependence, rationality, and animality, and his discussion of disability specifically, provides an occasion to further reflect upon status of the normal/abnormal binary and the goals of normalization.

There are many ways in which MacIntyre's book is sympathetic to the critique of the medical model of disability. First, he explicitly discusses the contextual nature of defining disability, thus recognizing the limits of a purely medical model of disability. Second, his articulation of the virtue of acknowledged dependence and his claim that we should "reassert our human animality" suggest that the attempt to draw a definitive, objective line between the "disabled" and the "non-disabled" is problematic, thus echoing Amundson's and Canguilhem's arguments against the objective ontological status of the "normal" and the "pathological/abnormal" as natural kinds. Finally, his account of individual and communal flourishing may provide an alternative to the language of normalization. However, the fact that MacIntyre's discussion of disability is firmly rooted in Aristotelian soil complicates the picture and makes his account potentially open to criticisms similar to those presented in the critiques of normalcy. While I regrettably cannot give this remarkable book the close attention it deserves here, I will briefly consider the place of cognitive disability in this work.

MacIntyre situates his book within contemporary philosophical discourse about disability, and offers a characterization of disability that is consonant with the social model of disability. He writes,

> It is and perhaps always has been a common assumption that blindness, deafness, deformed or injured limbs, and the like exclude the sufferer from more than a very, very limited set of possibilities. And this has often been treated as if it were a fact of nature. What is thereby obscured is the extent to which, whether, and how far the obstacles presented by those afflictions can be overcome or circumvented depends not only on the resources of the disabled but also on what others contribute. . . . What disability amounts to, that is, depends not just on the disabled individual but on the groups of which that individual is a member" (MacIntyre, 75).

The belief that disability cannot be reduced to an objectively "bad" condition located solely in the individual echoes the work of many disability theorists who challenge a purely medical model of disability. Yet MacIntyre takes his discussion further and challenges the ways philosophers have traditionally drawn a strict line between humans and nonhumans, a critique that suggests that certain conceptions of disability might also be revised.

MacIntyre focuses his book on the relationship between rationality, dependence, and animality, and urges us to resist the temptation to ignore, deny, or move beyond our fundamental animality: "We never completely transcend our animality."[21] In light of this, he hopes to "generally undermine the cultural influence of a picture of human nature according to which we are animals and in addition something else."[22] Thus he claims that there is a spectrum, rather than a strict binary, between the human and nonhuman. This implies that disability, vulnerability, impairment, and dependence are all part of the human experience and that there are particular virtues associated with acknowledging our dependence upon others: ". . . any adequate education into the virtues will be one that enables us to give their due to a set of virtues that are the necessary counterpart to the virtues of independence, the virtues of acknowledged dependence."[23] Rather than viewing the persons with disabilities as radically other by virtue of their dependence on others, MacIntyre suggests that all human beings begin life in a state of dependence, and, at various points, we all share in the experience of vulnerability. Thus, "Of the brain-damaged, of those almost incapable of movement, of the autistic, or all such we have to say: this could have been us. Their mischances could have been ours, our good fortune could have been theirs."[24] His claim that the line that separates the disabled from the nondisabled is somewhat arbitrary, permeable, and most importantly, one that traces back to each person's entrance into the world in a state of dependence, is an important step in addressing the oppression of persons with severe cognitive disabilities, many of whom remain in a state of dependence in ways that nondisabled persons do not.

However, as I have argued elsewhere, there may be reasons for being cautious about viewing persons with disabilities as mirrors for the nondisabled.[25] Furthermore, while the turn to our animality and our fundamental dependence on others as human beings is potentially liberatory for persons with disabilities, it is important to recognize that the philosophical attempt to "transcend our animality" of which MacIntyre is critical has not applied to all human beings. In fact, there is a long history of defining persons with cognitive disabilities (and other groups based on race, ethnicity, and gender, for example) in terms of their proximity to nonhuman animals. I think that arguments that juxtapose the "radically disabled" or the "mentally retarded" with discussions of animality should acknowledge the persistence, power (both discursive and concrete), and potentially deleterious effects of drawing a connection between cognitively disabled humans and nonhuman animals.

In considering the way this relationship has been forged in philosophical literature, I have found two types of arguments. Some focus on the mentally retarded to make claims about the moral status of animals, and others use the case of nonhuman animals to make statements about the moral status of the mentally retarded. In both cases, I have found that the link between animals and the "mentally retarded" in philosophical literature manifests itself in a number of ways. First, it can be *comparative*, whereby the condition or status of those labeled mentally retarded is compared with animals, or the relationship between "normal" human beings and the "mentally retarded" is thought analogous to our relationship with animals. For example, in "The Rights of the Retarded," Anthony Woozley draws a comparison between the "mentally retarded" and a dog to illustrate his point that they both lack a sense of justice: "A dog can look at you pleadingly, or even perhaps accusingly; but to say that he is pleading for justice, or accusing you of injustice, is to attribute to him a concept which it would be rash to suppose that he has; the same must be true of many of the retarded."[26] The relationship can also be *definitional*, where the "mentally retarded" are actually defined by their "animal-like" qualities or placed in the same moral category as nonhumans. Jeff McMahan concludes that in one sense, a distinction cannot be made between some disabled humans and nonhuman animals: "How a being ought to be treated depends, to some significant extent, on its intrinsic properties—in particular, its psychological properties and capacities. With respect to this dimension of morality, there is nothing to distinguish the cognitively impaired from comparably endowed nonhuman animals."[27]

There are questions regarding speciesism and definitions of personhood that time does not permit me to address here. However, it is important to note that the association between the cognitively disabled and nonhuman animals has not always been a positive one. Paul Spicker, in "Mental Handicap and Citizenship (1994)," argues that the comparison of the "mentally retarded" to animals is unfortunate for two reasons: first, "the moral rights accorded to humans and animals are not equivalent"; and second, "the behaviour of people toward animals is generally different from the behaviour of people toward other people. The identification of mentally handicapped people with animals is liable to change the way in which other people behave towards them."[28] In fact, the history of mental retardation bears out this second point. In many instances, appeals to the "animal-like qualities" of persons with cognitive disabilities were used to justify the horrific conditions and treatment to which they were subjected. Therefore, while MacIntyre's blurring of the human/nonhuman animal distinction allows us to rethink the virtue of independence in light of the virtue of acknowledged dependence, I would argue that discussions of human animality in the context of cognitive disability (and disability generally) must acknowledge the long legacy whereby human status has been denied to those who were believed to be incapable of transcending their animality.

There are ways in which MacIntyre's book supports challenges to the objective designation of disability as abnormal, particularly in light of his discussion of the ways in which all human beings experience dependence. However, I am left wondering to what extent the reification of the normal/abnormal boundary is implicit in his account of human flourishing, and what promise this concept may hold in addressing the distinct needs of persons with cognitive disabilities while avoiding the problems posed by the language of normalization.

MacIntyre states, "Whether or not a given individual or group is or is not flourishing qua member or members of whatever plant or animal species it is to which it or they belong is in itself a question of fact."[29] He goes on say he is committed to a somewhat naturalistic account "insofar as a plant or animal is flourishing, it is so in virtue of possessing some relevant set of natural characteristics."[30] The question, then, is what those relevant characteristics are for human beings. The following claim about human flourishing points to an answer: "The exercise of independent practical reasoning is one essential constituent to full human flourishing. It is not—as I have already insisted—that one cannot flourish at all, if unable to reason. Nonetheless, not to be able to reason soundly at the level of practice is a grave disability. It is also a defect not to be independent in one's reasoning."[31] Where does this portrait leave persons with cognitive disabilities of varying degrees? MacIntyre does not expand upon what a notion of human flourishing that did not involve independent practical reasoning would look like, but his work leaves room for further reflection on this question. Just as Amundson, Silvers, and others have challenged the objective designation of certain physical impairments or conditions as "objectively bad," we might consider whether the same challenge can be made to conceptions of human flourishing that view the relationship between rationality and flourishing qua human as an essential one.

MacIntyre does address persons with severe disabilities, referring to the "retarded," the "radically disabled," and those with "extreme disablement." However, there are some ambiguities regarding the group under discussion, ambiguities that I would argue should be resolved in order to avoid a prototype effect, whereby the "severely disabled" are classified as a unified, unproblematic group. In his criticism of some philosophical discussions of nonhuman animals, MacIntyre argues that philosophers often ignore important differences between nonhuman species.[32] Yet I would argue that MacIntyre falls prey to a similar tendency in discussing persons with cognitive disabilities in his work.

When speaking about the good parent, MacIntyre argues that "it is the parents of the seriously disabled who are the paradigms of good motherhood and fatherhood as such, who provide the model for and the key to the

work of all parents," and claims that the parents of "children who are severely disabled . . . have undertaken one of the most demanding kinds of
work that there is."[33] It is unclear how he would define the "seriously disabled," though the next paragraph argues that the function of a good parent is "to bring the child to the point at which it is educable, not only by
them but also by a variety of other different kinds of teacher. And this is the
first step towards making the child independent as a reasoner."[34] While the
first statement may well be true about parents of the severely disabled
being the paragon of parenthood, the goals of good parenthood as outlined
by MacIntyre are impossible for parents of children who can never become
independent reasoners.

MacIntyre rightly points out that "the radically disabled individual
needs someone who will speak for her or him as, so to speak, a second
self. . . . Yet no one will be able to speak adequately for me who does not already know me." Thus, our "second selves" are those who will know "how I
have judged my good in various situations in the past."[35] However, again
there is a question of which "radically disabled" cases he means. Insofar as
many of the severely cognitively impaired have never been capable of such
judgments, his characterization here would not apply. Thus, the question
remains what a concept of human flourishing that could accommodate
various degrees of cognitive disability would look like and what it would
mean to find a "second self" in these cases.

MacIntyre explains that the concept of flourishing is one and the same
for all species, even though "what it needs to flourish is to develop the distinctive powers that it possesses qua member of that species."[36] The question I am left with is whether, in defining a set of relevant natural characteristics, one is forced to admit the existence of an objective "norm" for
human beings, thereby reifying the normal/abnormal binary. Must the
concept of flourishing be grounded in biological fact, and what are the implications of such a definition given Canguilhem's and Amundson's critiques? The plausibility of MacIntyre's account of flourishing leads to further questions regarding cognitive disability, specifically. In claiming that
"it is not that one cannot flourish at all, if unable to reason," and in articulating the virtue of acknowledged dependence, MacIntyre's book leaves
opens the possibility of reimagining what human flourishing might mean
in individuals for whom independence of any kind is impossible.[37] Moreover, in light of the problems with programs that attempt to "normalize"
persons with cognitive disabilities, MacIntyre's discussion of communal
flourishing may provide a preferable alternative: "Those who benefit from
that communal flourishing will include those least capable of individual

practical reasoning . . . and their individual flourishing will be an important index of the whole community. For it is insofar as it is *need* that provides reason for action for the members of some particular community
that that community flourishes."[38] MacIntyre's book is a valuable addition
to philosophical literature on cognitive disability because, in reconfiguring
the relationship between rationality, dependence, and animality, he addresses three concepts that have been instrumental in the very definition of
mental retardation as a classification. In further exploring this relationship,
perhaps we can better understand the ways in which scientific and philosophical discourses have, in the name of these concepts, justified the oppressive practices towards persons with cognitive disabilities.

Conclusion

What does this analysis imply for the use of the abnormal/normal distinction and the goal of normalization in relation to persons with cognitive
disabilities? Should we retire the concepts of normalcy and normalization
completely? Unfortunately, I do not have a quick and ready answer. However, I will make three modest claims. First, engaging in a critique of the
fallacies, myths, and oppressive effects of the discourses of normalcy in
both the scientific and nonscientific contexts is crucial to achieving a
deeper understanding of the ways in which disability functions as a category. Second, for these critiques to be most effective a greater degree of
specificity is required. Here I have found the works of Michel Foucault
compelling because of their focus on the local and the particular. While
Foucault did discuss the broad class of individuals he referred to as "the ab-
normals," his work on the emergence of particular kinds of individuals
(e.g. the madman, the delinquent, the pervert) reminds one that multiple
factors and power relations can contribute to a particular group's relega-
tion to the "abnormal." In the spirit of a Foucauldian approach, then, I
maintain that the complexity of a category like mental retardation must be
addressed in critiques of normalcy and normalization, and in definitions of
human flourishing.

Finally, insofar as it is both philosophically and politically necessary to
disentangle the strands of normalcy discourse from characterizations of
disability and normalizing practices, I maintain that this task is more pressing than an attempt to do away with the categories of normal/abnormal altogether. I say this in part because I agree with Foucault when he describes
the nature of normalizing power in *Discipline and Punish* (1979): "The
judges of normality are present everywhere. We are in the society of the

teacher-judge, the doctor-judge, the educator-judge, the 'social worker'-judge; it is on them that the universal reign of the normative is based; and each individual, wherever he may find himself, subjects to it his body, his gestures, his behavior, his aptitudes, his achievements."[39] Whether or not the judges of normality will haunt a definition of human flourishing that can include persons with cognitive disabilities remains to be answered.

Notes

1. Lennard Davis, *Enforcing Normalcy: Disability, Deafness, and the Body* (London: Verso, 1995), 49.
2. Susan Wendell, *The Rejected Body: Feminist Philosophical Reflections on Disability* (New York: Routledge, 1996).
3. Anita Silvers, "Formal Justice," in *Disability, Difference, Discrimination* Anita Silvers, David Wasserman, and Mary Mahowald, eds. (Lanham, Md: Rowman & Littlefield Publishers, Inc., 1998), 74.
4. Richard Jenkins, "Culture, classification, and (in)competence," in *Questions of Competence: Culture, Classification and Intellectual Disability* Richard Jenkins, ed. (Cambridge, U.K.: Cambridge University Press, 1998), 22.
5. George Canguilhem, *The Normal and the Pathological*, trans. Carolyn Fawcett (New York: Zone Books, 1991), 226.
6. Ibid., 136–7.
7. Ron Amundson, "Biological Normality and the ADA" in *Americans With Disabilities: Exploring Implications of the Law for Individuals and Institutions* Leslie Pickering Francis and Anita Silvers, eds. (New York: Routledge, 2000) 103.
8. Ibid., 103.
9. Ibid.
10. I owe this insight to an anonymous reviewer of an earlier draft, who pointed out that many different sciences go to great lengths to reject the idea of "abnormal" as having only statistical meaning.
11. I must confess that I have avoided the language of social construction in discussing the nature of this classification, in part because I find it too broad to address the complexities of this category. However, I find Ian Hacking's clarifications regarding the uses and meanings of this phrase helpful in various ways in which we might say that mental retardation is "socially constructed." See Ian Hacking, *The Social Construction of What?* (Cambridge, Mass.: Harvard University Press, 1999).
12. See Licia Carlson, "Docile Bodies, Docile Minds: Foucauldian Reflections on Mental Retardation," forthcoming, in *Foucault and the Government of Disability*, Shelley Tremain, ed. (New York: Routledge).
13. Simi Linton, for example, points out that the terms "normal" and "abnormal" are used in multiple disciplines in the social sciences, wherein the statistical, value-laden nature of these designations are overshadowed, and the category of abnormal intelligence and/or behavior becomes reified. She goes on to argue that the ways in which treating normal/abnormal as absolute categories has profound consequences for social structures, individual decisions, and curricu-

lar structures. See *Claiming Disability: Knowledge and Identity* (New York: New York University Press, 1998), 22–25.

14. George Lakoff, *Women, Fire, and Dangerous Things: What Categories Reveal About the Mind* (Chicago: University of Chicago Press, 1987), 41.

15. Perhaps the most clear historical example is the "moron" who, in the early decades of the twentieth century, became *the* symbol of feeblemindedness. By virtue of the ability to "pass" as normal, the "moron" was considered the most dangerous kind of feebleminded individual, and because of the success of IQ tests, it was possible to definitively pick out this new "type" of individual. See James Trent, *Inventing the Feeble Mind: A History of Mental Retardation in the United States* (Berkeley: University of California Press, 1994), chs. 5 and 6. For a striking philosophical example of this, see Peter Singer, *Animal Liberation* (London: Pimlico, 1995).

16. Wolf Wolfensberger, "Social role valorization: A Proposed new term for the principle of normalization," *Mental Retardation*, 21: 234–9.

17. See Linton 1998, Jenkins 1998.

18. Eva Feder Kittay, "At Home With My Daughter," in Leslie Pickering Francis and Anita Silvers, eds., *Americans with Disabilities: Exploring Implications of the Law for Individuals and Institutions* (New York: Routledge, 2000), 74.

19. Jenkins, 1998, 21–22.

20. At the same time, there are important ways in which people with cognitive disabilities have expressed the desire to engage in "normal" behavior and to be viewed as "normal people," suggesting that the force of this category cannot be overlooked or taken for granted by those of us who may enjoy the benefits of conforming to these norms. For an interesting discussion of these dynamics, see Michael F. Angrosino, "Mental disability in the United States," in *Questions of Competence*, 25–53.

21. MacIntyre, 8.

22. Ibid., 50.

23. Ibid., 120.

24. Ibid., 101.

25. See Licia Carlson, "Cognitive Ableism and Disability Studies: Feminist Reflections on the History of Mental Retardation," *Hypatia* 16(4): 124–146, 142; Linton, 14.

26. Anthony Woozley, "The Rights of the Retarded," in *Ethics and Mental Retardation* Loretta Kopelman and J.C. Moskop, eds. (Dordrecht, Holland: D. Reidel Publishing Company, 1984), 51. He explains who the "many" are: "A person does not lack a sense of justice by being retarded- unless he is retarded enough; and surely plenty are retarded enough for that . . . he does not have the comprehension which you must have to have a sense of justice." (50–1)

27. Jeff McMahan, "Cognitive Disability, Misfortune, and Justice," *Philosophy and Public Affairs* 25 (1): 3–35, 32.

28. Paul Spiker, "Mental Handicap and Citizenship," *Journal of Applied Philosophy* 7 (2): 142.

29. MacIntyre, 64–5.

30. Ibid., 78.

31. Ibid., 105.

32. Ibid., 48.
33. Ibid., 91.
34. Ibid.
35. Ibid., 139.
36. Ibid., 64.
37. I think Eva Feder Kittay's work is an important contribution here. See *Love's Labor: Essays on Equality, Dependence and Care*, (London and New York: Routledge, 1999); "At Home with My Daughter," in Francis and Silvers, *Americans with Disabilities*.
38. MacIntyre, 108–9.
39. Michel Foucault, *Discipline and Punish: The Birth of the Prison*, trans. Alan Sheridan (New York: Vintage Books, 1979) 304.

References

Amundson, Ron. "Biological Normality and the ADA." In *Americans With Disabilities: Exploring Implications of the Law for Individuals and Institutions*, Leslie Pickering Francis and Anita Silvers, eds. New York: Routledge, 2000.

Angrosino, Michael F. "Mental Disability in the United States: An Interactionist Perspective." in *Questions of Competence*, Richard Jenkins, ed. Cambridge, U.K.: Cambridge University Press, 1998, 25–53.

Canguilhem, George. *The Normal and the Pathological*, trans. Carolyn Fawcett (New York: Zone Books, 1991).

Carlson, Licia. "Cognitive Ableism and Disability Studies: Feminist Reflections on the History of Mental Retardation." *Hypatia* 16(4): 124–146.

———. "Docile Bodies, Docile Minds: Foucauldian Reflections on Mental Retardation." In *Foucault and the Government of Disability*, Shelley Tremain, ed. New York: Routledge, forthcoming.

Davis, Lennard J. *Enforcing Normalcy: Disability, Deafness, and the Body*. London: Verso, 1995.

Foucault, Michel. *Discipline and Punish: The Birth of the Prison*, trans. Alan Sheridan. New York: Vintage Books, 1979.

Hacking, Ian. *The Social Construction of What?* Cambridge, Mass.: Harvard University Press, 1999.

Jenkins, Richard. "Culture, classification, and (in)competence." In *Questions of Competence: Culture, Classification and Intellectual Disability*, Richard Jenkins, ed. Cambridge, U.K.: Cambridge University Press, 1998.

Johnson, Kelley. *Deinstitutionalizing Women: An Ethnographic Study of Instituional Closure*. Cambridge, U.K.: Cambridge University Press, 1998.

Kittay, Eva Feder. "At Home With My Daughter Sesha." In *Americans with Disabilities;* Leslie Pickering Francis and Anita Silvers, eds. New York: Routledge, 2000.

———. *Love's Labor: Essays on Equality, Dependence and Care*. London and New York: Routledge, 1999.

Lakoff, George. *Women, Fire, and Dangerous Things: What Categories Reveal About the Mind.* Chicago: University of Chicago Press, 1987.

Linton, Simi. *Claiming Disability: Knowledge and Identity.* New York University Press, 1998.

MacIntyre, Alisdair. *Dependent Rational Animals.* Chicago: Open Court Publishing Co, 1999.

McMahan, Jeff. "Cognitive Disability, Misfortune, and Justice." *Philosophy and Public Affairs* 25 (1): 3–35.

Silvers, Anita, "Formal Justice." In *Disability, Difference, Discrimination,* Anita Silvers, David Wasserman, Mary Mahowald, eds. Lanham, Md.: Rowman and Littlefield Publishers, Inc., 1998.

Singer, Peter. *Animal Liberation.* London: Pimlico, 1995.

Spicker, Paul. "Mental Handicap and Citizenship." *Journal of Applied Philosophy* 7 (2): 139, 1990.

Trent, James. *Inventing the Feeble Mind: A History of Mental Retardation in the United States.* Berkeley: University of California Press, 1994.

Wendell, Susan. *The Rejected Body: Feminist Philosophical Reflections on Disability.* New York: Routledge, 1996.

Wolfensberger, Wolf. "Social Role Valorization: A Proposed New Term for the Principle of Normalization." *Mental Retardation* 21: 234–9, 1983.

Woozley, Anthony. "The Rights of the Retarded." In *Ethics and Mental Retardation,* Loretta Kopelman and J.C. Moskop, eds. Dordrecht, Holland: D. Reidel Publishing Company, 1984.

Conformity through Cosmetic Surgery: The Medical Erasure of Race and Disability

Sara Goering

Individuals conform to many social norms, including those related to the body and its appearance. We style our hair, put on make-up, and do painstaking exercises to create the "right" bodily appearance, all without moral suspicion.[1] We tend to believe, instead, that such decisions are personal, affect only the individual, and are therefore immune from public moral condemnation.[2] But some individuals take more drastic steps to gain conformity—they have their bodies surgically altered to fit the norms of appearance. Proponents of cosmetic surgery may argue that it is merely an extension of other beauty practices, and therefore morally unproblematic. In this paper I will argue that it is, instead, a pernicious practice that threatens diversity.[3] My argument will focus on the connections between cosmetic surgery and the medical establishment, showing how the model of objective treatment is used to justify such surgeries. I will show how cosmetic surgery has been used to erase the physical signs of (non-white) race[4] and of mental retardation/disability[5], often at the urging of surgeons who purport to be using an objective model of health to identify problematic features of appearance. Finally, I suggest that this conflation of health and socially acceptable appearance not only threatens diversity through the practice of cosmetic surgery, but may also usher in a form of "backdoor eugenics"[6] as we advance our genetic technologies. In other words, we may eliminate nondominant group traits via many instances of rational individual choice, without recognizing the broader effect such actions will have on society as whole.

Recent academic works on the history of plastic surgery[7] highlight the longstanding tension between plastic surgery for treatment/reconstruction and for aesthetic enhancement. Historically, surgeons who advanced a medical agenda fought to distinguish themselves from surgeons who used the same skills for nonmedical, cosmetic purposes. The former saw themselves as healers who used surgery to relieve pain, treat deformity, and restore function. Their aim was good health and normal functioning, and they offered realistic expectations for improvement, cautioning their patients about the physical risks of surgery. They distinguished themselves from mere "quacks" who offered surgery with the aim of profit, promising perfection and advertising widely to take advantage of vanity.[8] The "serious" plastic surgeons considered cosmetic uses of their skills to be in conflict with the basic medical principle "do no harm."[9] Individuals who requested purely cosmetic surgeries were not, after all, physically unhealthy. In the view of the early plastic surgery societies, surgeons who promoted themselves and took on frivolous cosmetic cases were considered to be making a mockery of the medical profession.[10] Thus, the mainstream surgeons recognized the need to draw a moral line between reconstructive and cosmetic surgery. However, even in the early days, this line was difficult to define.

The line between reconstructive and cosmetic surgery was fuzzy for several reasons. First, some surgeons came to recognize that facial imperfections, even those which did not adversely affect physical functioning, were most visible, and consequently most socially damaging.[11] A man whose nose was large, red, and bulbous was perceived as a drunk, compromising his earning potential. Early twentieth century plastic surgery societies thus agreed to endorse cosmetic facial surgery when necessary to ensure a man's ability to earn a living.[12] They had, after all, treated World War I soldiers with battle-induced facial deformities, in order to help them return to civilian life without ostracism. But if similar social rejection was dealt a man who had not been to war, who had instead received facial imperfections through unlucky genetics or congenital abnormalities, then fairness seemed to require the treatment of him as well.

According to Elizabeth Haiken's historical analysis of cosmetic surgery, this blurring of the line between reconstructive and cosmetic surgery was made even worse with the introduction of Adler's inferiority complex, which was used to connect poor mental health with deformity (or even perceived deformity).[13] "As physical conditions ranging from deformity to simple ugliness were defined as the cause of an inferiority complex and thus equated with mental illness, cosmetic surgery, once associated with lack of mental balance and overweening vanity, came to be seen as a step toward mental health."[14] This shift, then, suggested that medicine's proper role was the treatment of suffering related to bodily conditions, even if what created the suffering on account of those bodily conditions happened to be societal norms rather than intrinsic defects.[15] An individual's perception that her looks were inferior created suffering, which doctors could relieve. Although surgery or psychotherapy might successfully rid the individual of her suffering, surgery was, and is, often considered to be less costly and more efficient.[16] Thus, "serious" plastic surgery expanded to include surgery for reconstruction or treatment of the psyche.[17] Sander Gilman argues that this shift helped to reestablish cosmetic surgery as a reputable practice in the medical community and in the public eye. "Medicine's job became correcting the appearance of illness as well as its pathology."[18]

Of course, one problem with this shift is that what constitutes the "appearance of illness" depends in large part upon what the societal standards of disease/illness, health, and appearance are. In philosophy of medicine, a disease is typically defined as "an adverse departure from normal species functioning."[19] Thus, in order to determine if something is a disease, we must know the statistical range of functioning for the species, determine whether or not a particular deviation falls outside of the normal range, and assess whether the deviation is positive or negative.[20] While this model appears to work for many prototypical diseases (e.g., diabetes, Tay-Sachs, etc.), looking at less "clean" cases highlights its problematic grounding. On the one hand, just because something is statistically normal does not mean that we necessarily value it, or ought to value it. For instance, aging women's decreasing bone density is statistically normal, but not something we prefer to maintain. On the other hand, we could also say that some statistically abnormal conditions (e.g., deafness) need not be devalued simply because they are not the norm. Indeed, individuals who live with such conditions often find the majority's treatment of them as "defective" frustrating and unjustified. Members of the Deaf culture "characteristically think it is a good thing to be Deaf and would like to see more of it."[21] As Anita Silvers has pointed out, the standard definition of disease unfairly "valorizes the normal"[22] without explaining why what is statistically normal is also normatively valuable. Recent works have criticized the definition of health as normal species functioning,[23] yet this conceptual understanding still maintains a place of prominence in bioethics.[24]

The use of the species-typical functioning model is particularly disturbing when we look at medicine as applied to bodily appearance. Health, as regards appearance, is defined objectively. Plastic surgeons have historically referred to measurement charts that purportedly describe the best bodily ratios (e.g., nose to face length, distance between the eyes, etc.).[25]

Such idealized "maps" of the human body lend the practice of altering the body a certain authority. The cosmetic surgeon becomes the expert on bodily proportions, just as the endocrinologist is the expert on hormonal balances. This allows the surgeon to offer advice to the patient, not only about what she desires, but about what she ought to desire (to improve her body's appearance in an objective manner). Patients' desires are thus sometimes "created" in the surgeon's office; they are encouraged to pursue surgery for "faults" they did not previously recognize in themselves.[26]

Just as we question the "valorization of the normal" regarding functioning, so too should we question the "normal" appearance standard that guides the cosmetic surgeons in their practice. Who defines the model? Although we might be tempted to think of the model as a mere statistical mean, what we find in practice are not attempts to look average. Rather, we find cosmetic surgeons and their patients alike seeking to create an idealized body. It is an odd slide, from health as normal species functioning, to health as normal species appearance (via mental health), to health as idealized species appearance.

To better understand the ethical ramifications of this shift, I propose to look at two bodily conditions that have historically been "treated" through cosmetic surgery, and continue to be so treated today (in increasing numbers).[27] Both are defined as atypical by suspect norms of appearance: nonwhite facial features, and facial features associated with mental retardation.

Race and Cosmetic Facial Surgery

Stereotypical nonwhite racial features[28] (e.g., broad or hooked noses, noncreased eyelids, etc.) have historically been designated as "ugly" or at least suboptimal by a racist American society.[29] Thus, individuals who partake in cosmetic surgery to erase the phenotypic signs of race often claim simply to be attempting to become normal or beautiful. But our societal images of beauty are deeply engrained in our discriminatory history, even if individuals often fail to make the connection. Margaret Little notes, "It is *no accident* that the standard of beauty prevalent in the West favors white European features over black African ones."[30] Whiteness has historically been valorized in science and religion, leading to racist conceptions of natural hierarchies, which view whites as more fully developed than other races. In addition, we often wrongly assume that "bodily beauty and deformity covary with moral beauty and deformity as well as with general cultural and intellectual capacity."[31]

As a result, many individuals (both white and nonwhite) come to view racialized features as signs of ugliness. As Paul Taylor notes, "The modern black experience has been intimately bound up with a struggle against the cultural imperative to internalize the judgment of one's own thoroughgoing ugliness."[32] A whole social movement (from Toni Morrison in literature to Spike Lee in film to bell hooks and Cornel West in the academy) aims at overcoming these racist aesthetic standards. Yet they are difficult to avoid. These beliefs about appearance and character are internalized by nonwhites, as well as their white counterparts, and by those who work in the medical establishment, as well as those who merely make use of it.[33] Such tendencies can result in significant discrimination against people of color, making "passing" as white appear to be a rational goal. The consequence is a wide variety of racialized beautification practices (e.g., skin bleaching, hair straightening, etc.) as well as intrusive surgical "solutions" (e.g., nose straightening for Jewish noses, narrowing for African noses, implanting of cartilage for Asian noses, eyelid "corrections" for Asian eyes, etc.).

Although many women believe such beauty practices to be simply playful or reflective of arbitrary personal preferences,[34] the fact that most of the practices aim at one specific kind of body suggests that much more than personal preference is at issue. Susan Bordo notes, "When we look at the pursuit of beauty as a normalizing discipline, it becomes clear that not all body transformations are the same."[35] People get colored contacts to make brown eyes blue or green, but rarely the reverse. People straighten their noses with surgery but do not ask for hooks. Thus, women's responses to the pressures for racial normalization are not arbitrary and are probably not best described as playful either, given that significant consequences ride on one's ability to meet the racialized norm.

Eugenia Kaw studied Asian-American women's attempts to "escape persisting racial prejudice that correlates their stereotyped genetic physical features . . . with negative behavioral characteristics, such as passivity, dullness, and a lack of sociability."[36] Kaw found that these women claimed to be "proud to be Asian American," but they sought to Westernize their appearances in hopes of gaining social standing. They recognized the market value of a Westernized appearance; indeed, they may have themselves learned to view it as more beautiful and therefore rationally requested surgery as a way to improve their individual status. Although we might wonder about the motivations of the women themselves, they were no doubt also influenced by surgeons who encouraged them to undergo the knife. As Kaw notes, "Through the subtle and often unconscious manipulation of racial and gender ideologies, medicine, as a producer of norms, and the larger consumer society of which it is a part, encourage Asian American women to mutilate their bodies to conform to an ethnocentric norm."[37] The surgeons assume, perhaps unwittingly, a Western norm of beauty and then discuss Asian American features as inadequate ("using words like 'without' 'lack of' 'flat' 'full' and 'sleepy' "[38]) by comparison to that norm. The surgeons and women note that they just want a "more open appearance" to their eyes, but fail to see that this very desire may be due to racial prejudice. Eye shapes, so long as they do not impede function (e.g., droopy eyelids that decrease the field of vision), could be seen as a valuable, interesting form of human diversity. Why valorize a single eye shape as the ideal? In a more appearance-tolerant society, we might expect to find all kinds of facial structures, both as a result of natural causes and from truly "playful" elective surgeries. As Kaw notes, "If the types of cosmetic surgery Asian Americans opt for are truly individual choices, one would expect to see a number of Asians who admire and desire eyes without a crease or a nose without a bridge. Yet the doctors can refer to no cases involving Asian Americans who wanted to get rid of their creases or who wanted to flatten their noses."[39]

The strength and extent of this racialized beauty norm may be growing. Elizabeth Haiken notes that what counts as the ideal face has changed with the passage of time, but "what we see when we look at popular culture is both more inclusive than ever before and yet frustratingly narrow. And if this is not enough to convince, a quick tour around the globe—from China, where a mark of the new bourgeoisie is the 'Big Nose' to Lima, Peru, where to have a 'Hispanic' nose (rather than Indian) is to move up a social class or more and cut-rate surgical clinics are springing up by the dozens to meet this new need—testifies to the enduring power of whiteness."[40]

Cosmetic surgeons who claim to be treating the mental health of women[41] in an objective fashion use the authority of their positions within the medical establishment to encourage such racialized transformations. Although individual women who pursue this strategy may well be acting on rational grounds vis-à-vis their individual situations (i.e., in a racist society, individual benefits in social and work situations may well follow such transformations), the commonality of this practice taken in the aggregate suggests serious problems. The field of cosmetic surgery's promise to help individuals achieve good mental health through such transformations ignores the background racist conditions, and indeed reinforces those conditions through its practice.

Mental Retardation and Cosmetic Facial Surgery

In the mid-1970s, cosmetic surgery was used on children with Down syndrome with the aim of improving their quality of life by removing the stigmatized facial features associated with mental retardation (e.g., "lifting the sunken bridge of the nose, effacing the epicanthic fold, correcting the oblique lid axis, augmenting the chin," etc.[42]). This practice raised significant controversy, with some professionals arguing that it helps to integrate persons with Down syndrome into society, and others claiming that it is not effective and may even cause psychological problems.[43] Although one of the purported goals of such surgeries is improved physical functioning (e.g., decreasing tongue size to improve the individual's ability to breathe, eat, and speak clearly), effectiveness studies suggest that physical functioning is not significantly altered with most such surgeries.[44] Appearance improvement is thus a primary aim, with a secondary hope for increased social acceptance. Studies of appearance judgments, however, generally find that while surgeons and parents of the patients noticed improvement, nonrelated individuals (teachers, peer-age kids, etc.) saw no difference pre- and postoperatively.[45] As Katz and Kravetz note, "The lack of strong evidence for the operation's effectiveness . . . reveals the degree to which the professional decision to recommend the operation and the parents' decision to . . . accept this recommendation is more a matter of value, or hope, than of fact. When making these decisions . . . these professionals and parents are to some extent answering the questions as to who should bear the burden of change, the stigmatized individual or the stigmatizing society."[46]

Regardless of the effectiveness of the surgery, we might rightly ask whether the decision to alter the facial features of the child with Down syndrome ought to occur. While there is no question that individuals with Down syndrome often suffer from teasing and negative evaluations based on their appearance,[47] it is less clear what motivates this treatment and what would best resolve the problem. The teasing may not be on account of the appearance itself so much as its association with behavioral traits that often coincide with mental retardation.[48] In this case, altering the physical features might result in less teasing from a distance, but may not have any affect on social acceptance "up close." Also, in some cases, the external sign of mental retardation may in fact be advantageous to the individual. As Dodd and Leahy note, "In practice, there is some advantage in being recognizably disabled. Severely hearing impaired individuals, for example, know that one of the problems associated with an invisible handicap is that no allowance is made for them."[49] Thus, the surgery undertaken in the name of improving the prospects of the individual with mental retardation may, in fact, if successful, harm that person by setting up false expectations in the eyes of the "normal" public. Of course, advocates of the surgery might argue that it is just these unfairly lowered expectations for Down syndrome children that keep them from reaching their full potential. One mother who opted for the surgery for her son says, "What I'm hoping for is that if Steven is not readily identifiable, people will expect more of him. The more expectations people put on him, the better he performs."[50] However, a more direct approach to remedying the situation clearly would involve addressing these expectations directly, rather than trying to circumvent them through cosmetic surgery. That is, educational mainstreaming of children with Down syndrome and general education about the range of abilities of individuals with Down syndrome may help to raise performance expectations without requiring such surgeries. At the same time, some critics say, "The surgery is itself a kind of rejection, a message that the children are not acceptable as themselves. It is society's preoccupation with 'good looks' that should change, these critics argue, not the faces of Down children."[51]

Unfortunately, the matter is much more complicated than it at first appears. A look at the history of other appearance-oriented medical alterations for children offers numerous precedents for intervention. If appearance can in fact be improved (by the standards of nonrelated observers),[52] then a surgeon might simply point to other unnecessary and sometimes painful medical interventions that aim at improved appearance and are widely accepted across society. Strauss notes, "If we permit children with minor physical deviances to have orthodontic treatment, why should this be denied to those with more severe defects?"[53] R.B. Jones adds, "[A] parallel can be drawn with another form of surgery undertaken in order that the child may conform to the norms of the society in which he or she lives: circumcision."[54] Neither male nor female circumcision is usually undertaken for health or functional purposes. Yet, in most Westernized nations, we allow and commonly practice male circumcision, while outlawing female circumcision as constitutive of abuse of a child.[55] This distinction clearly deserves attention itself. However, if solid, it might set a precedent for denying certain kinds of nontherapeutic physical interventions to children (based on their health risks, potential psychological damage, etc.). Jones suggests that perhaps Down's surgery[56] should also be outlawed since it is "equally major, painful and always therapeutically unnecessary."[57]

In addition to concerns over the pain and suffering that may accompany cosmetic surgeries for Down's children, we must also recognize how such treatment will affect and reinforce the underlying norms that lead parents to consider such surgery in the first place. This takes several forms. First, what is possible within the medical arena helps to determine what is considered standard practice. "The possibility of restoring the face has made an unrestored face less acceptable. For example, what was acceptably 'bucktoothed' in the 1950s has now become a 'dentofacial deformity' and is treated with surgery and orthodontics."[58] What is normal is defined in part by what is possible (rather than what is statistically average). Furthermore, in taking the option of altering the face to meet appearance norms, we assimilate to them, and in so doing, we exacerbate the pressures to meet that norm.[59] We make the "other" more apparent and abnormal by masking our own difference.

Cosmetic Surgery, Morality, and Public Policy

Having explored briefly the use of cosmetic surgery to erase or reduce the physical signs of nonwhite race and mental retardation/disability, what might we say about the ethical justifications of such practices? I doubt that many individuals who are aware of the practices find them morally untroubling. These surgeries are unnecessary, they involve the significant health risks that accompany any major surgery, and they appear to be motivated by the desire to fit in with suspect norms of appearance that unfairly denigrate the features characteristic of historically disenfranchised groups. And yet, individuals ask for them, and we are inclined to recognize individual prerogatives in making their own choices, especially about their own bodies.[60] What, then, can be said in response to the person who recognizes the troubling nature of such surgeries yet resists a negative moral or political evaluation of the practice? Such a person is likely to make one of three claims: 1) these surgeries are acceptable because they are autonomously chosen; 2) these surgeries are acceptable because they are innocuous, even if they arise in part from suspect social norms or lack full autonomy; or 3) these surgeries are acceptable because they are intended to create beauty rather than to erase the signs of race/disability, and the merely foreseen side effects of erasing the physical signs of race/disability are not morally significant.

Although many of the individuals who procure these surgeries may, in fact, believe that their choices are purely autonomous ("I'm doing it, for me"), such claims fail to investigate how the desires for surgery are formed and how the decisions to take action are made. The mere claim to autonomy or appearance of it does not make it so. Coercion and hypnotism clearly undermine autonomy, despite apparent voluntary action. Recent theories of relational autonomy suggest that other factors (e.g., diminished self worth,[61] poor development of competency for self-direction, self-discovery, or self-definition[62]) may constrain autonomy enough to make it questionable as a basis for informed consent and moral responsibility. Such theorists suggest that an adequate theory of (relational) autonomy must account for "the impact of social and political structures, especially sexism and other forms of oppression, on the lives and opportunities of individuals."[63] If we view individuals not as isolated, discrete entities, but rather as complex, interconnected entities, constituted in part through their relationships and by their culture, then we will have to reevaluate our judgments about autonomy. A woman who seeks cosmetic surgery for herself need not be coerced (e.g., by her surgeon or her spouse) in order to lack full autonomy in choosing it. The background norms of the society, which are racist, sexist, and ableist, may either constrain her autonomy through their constitutive effects on her social self, or through their causal limiting effect on her development of the skills of, for example, self-reflection and values questioning. Thus, despite some feminist work suggesting cosmetic surgery as an agency-empowering practice for women (one that avoids the "cultural dope" model of women coerced into cosmetic surgery by social forces out of their control),[64] we should not ignore the diminished autonomy that often results from oppressive background conditions and cultural values. They put limits on what we perceive as reasonable desires and realistic options to address our needs. To reconcile autonomy with the pressures of such influences, we have to look at the history and development of the individual and the relationships of power and social oppression that both give rise to cultural values and shape individual identities.[65]

Even if my critic accepts this first point, she might still claim that individual decisions for cosmetic surgery are, even if less than fully autonomous, innocuous in their effects. Here, it appears that my critic focuses exclusively on harms to the individual, rather than investigating possible harms to society. For the individual who undergoes surgery, the results may be excellent. She may find herself more easily accepted, more marketable, and generally with an advantage relative to other women. But the broader social effects may be quite negative. The "success" of the surgery reinforces the norm that encouraged it. As Sherwin notes, "With each use, it becomes ever more established as part of routine . . . care and thus helps consolidate the significant power and control that attaches to those who provide services deemed essential elements of good . . . care."[66] Granted, Sherwin was writing about the now-routine use of ultrasound technology in prenatal care, but the general effect is the same. As more individuals get cosmetic surgery, having it becomes part of the routine of aging, and the choice not to have it becomes less realistic. This affects not only women who choose not to have it at all, but also women who have it once and then realize the effects are not lasting. Repeat customers may become the norm.[67] Not only does this effect mean that individuals will experience more pressure to have cosmetic surgery, but it also suggests that available funds will be spent on surgical adaptation of appearance rather than other pressing social and ethical issues. We pay a price for the escalation of cosmetic surgery, even if we are inclined not to attend to it.

Finally, when the practice is so clearly directed to one racialized idea of beauty, and when people with disabilities are encouraged to undergo surgery so that they appear more normal, we must question the biases in our standards. Although my critic might suggest that cosmetic surgeons and their clients really are trying to improve the aesthetic value of the world, this line of argumentation is weak. One might suppose that the ideal of facial beauty converges on one look because it *is* objectively more beautiful, or perhaps more symmetric or proportional. Yet a closer look at the reasons why individuals have the surgeries suggests otherwise. Haiken cites a surgeon who notes, "I have operated on many persons of different ethnic groups who doubtless would never have considered surgical correction . . . if they had remained within their native borders."[68] This hesitation is not due to availability of surgical procedures in the home country, but rather to a different understanding of beauty and what is acceptably normal. In Sweden, Middle Eastern immigrants are seventeen times more likely to have the aesthetic rhinoplasty than ethnic Swedes, based on what the surgeons surmise are "assimilation difficulties" and "low tolerance of the society in accepting people with a foreign look."[69] The "foreign" look need not be asymmetrical; it is enough that it is not the norm. Much more work must be done to show that the standard of Western beauty is not grounded in an objective assessment of human beauty. Other works have investigated that topic,[70] and I leave it to future research to solidify these brief suggestions.

As a final note, this debate has significance for much more than the arena of cosmetic surgery. As we advance in our understanding of genetics and develop genetic therapies for ourselves and our future children, we will attempt to develop ethical practices based on established rules of medical intervention.[71] If we do not look carefully at our current uses of medical technology to enable us to meet appearance norms, we may find ourselves in a world in which we feel pressured to design all children to meet the Western standard of beauty. Each set of parents may make the sensible decision (from a private perspective) to give their child the best possible chance for success in a world biased against non-Western appearance. If that comes to be, one beautiful form of diversity—the vast human tapestry of color, shape, and unique proportions of faces—may be lost. We each ought to think beyond our individual best interest to see how, together, we can preserve the beauty of humanity.

Notes

1. Not all would agree. See Naomi Wolf, *The Beauty Myth: How Images of Beauty Are Used Against Women* (New York: Doubleday/Anchor Books, 1991); also Kirsten Dellinger and Christine Williams, "Makeup at Work: Negotiating Appearance Rules in the Workplace," *Gender & Society* 11(2) 1997: 151–77.

2. Following the harm principle of J.S. Mill, *On Liberty* (Indianapolis, Ind.: Hackett Publishing, 1978 [1859]).

3. I do not claim that it is necessarily a practice that threatens diversity, but only that its current use has that consequence. For a discussion of possible radical uses of cosmetic surgery, see Kathryn Morgan, "Women and the Knife: Cosmetic Surgery and the Colonization of Women's Bodies," in *Sex/Machine*, ed. Patrick Hopkins (Bloomington, Ind.: Indiana University Press, 1998); and Peg Brand, "Bound to Beauty: An Interview with Orlan," in *Beauty Matters*, ed. Peg Brand (Bloomington, Ind.: Indiana University Press, 2000).

4. Although the concept of race as a distinct biological entity has been widely denounced in recent years, its power as a tool of social organizing and grouping—for good and for ill—remains steadfast (see, e.g., Troy Duster, "The Sociology of Science and the Revolution in Molecular Biology" in the *Blackwell Companion to Sociology*, ed. Judith Blau (Oxford, U.K.: Blackwell Publishers, 2001); Angelo Corlett, "Latino Identity," *Public Affairs Quarterly* 13(3) (1999): 273–95. Individuals continue to act according to social notions of race (however fuzzy and inaccurate those notions are). Troy Duster raises concern over the possibility of a racial eugenics of the future, based on: 1) inaccurate but deeply entrenched public beliefs about the link between genes and race; 2) continuing racism and racial discrimination; and 3) a medical profession that acts from a slippery definition of health with a myopic focus on patient autonomy (Troy Duster, *Backdoor to Eugenics*, New York: Routledge, 1990). For the purposes of this paper, race need not be a genetic or biological fact, but only a socially significant category.

5. I use the broad term disability because cosmetic surgeries have been used to disguise or erase a broad range of disabilities. In this paper, however, I concentrate on cosmetic surgeries to eliminate the facial features characteristic of Down syndrome.

6. Duster, 1990.

7. Sander Gilman, *Making the Body Beautiful: A Cultural History of Aesthetic Surgery* (Princeton, N.J.: Princeton University Press, 1999); ———, *Creating Beauty to Cure the Soul: Race and Psychology in the Shaping of Aesthetic Surgery* (Durham, N.C.: Duke University Press, 1998); Elizabeth Haiken, *Venus Envy: A History of Cosmetic Surgery* (Baltimore, Md.: Johns Hopkins University Press, 1997).

8. See Haiken, 54.

9. For a larger discussion of this justification, see Haiken, 163.

10. A similar argument has been advanced against contemporary cosmetic surgeons in Franklin Miller, Howard Brody, and Kevin Chung "Cosmetic Surgery and the Internal Morality of Medicine," *Cambridge Quarterly of Healthcare Ethics* (9) 2000: 353–64.

11. Frances Cooke MacGregor notes "There is a vast difference between the kinds of social and psychological problems associated with physical handicaps (such as a paralyzed leg or the loss of an arm) and those stemming from deviations or distortions of the face (the symbol of one's identity and persona)—which evoke adverse reaction and social rejection." (225) "A Response to the 'Quasimodo Complex,'" *Journal of Clinical Ethics* 1(3) 1990: 224–6

12. As Haiken notes (38–39), this is a sexist distinction to make. They refused female requests for cosmetic surgery as mere vanity, but were willing to perform similar surgeries on men in the name of keeping them in the work world.

13. For an excellent discussion of this turn of events, see Haiken, ch. 3, "Consumer Culture and the Inferiority Complex," 91–130; and Gilman, ch. 6 "Alfred Adler's Inferiority Complex," in *Creating Beauty to Cure the Soul*, 100–10, 1998.

14. Haiken, 129. Indeed, this link between facial/body normality and good mental health/good life prospects was so strong that in 1927 the San Quentin prison allowed a project that tried to turn prisoners into "new men" through cosmetic surgery (Haiken, 109).

15. Deborah Sullivan refers to this as "scalpel psychiatry" in *Cosmetic Surgery: The Cutting Edge of Commercial Medicine in America* (New Brunswick, N.J.: Rutgers University Press, 2001), especially, ch. 3.

16. Similar claims have been made regarding the use of antidepressants, e.g., Prozac, vs. psychotherapy to address depression. See Carl Elliott, "Pursued by Happiness and Beaten Senseless: Prozac and the American Dream," *Hastings Center Report* 30(2) 2000: 7–12.

17. See Gilman, *Creating Beauty to Cure the Soul*, 5.

18. Gilman, 16.

19. Quote from Allen Buchanan, Dan Brock, Norman Daniels, and Daniel Wikler, *From Chance to Choice: Genetics and Justice* (Cambridge: Cambridge University Press, 2000) p. 72. For the history of this designation, see the work of Christopher Boorse "On the Distinction between Disease and Illness" in *Concepts of Health and Disease* (eds., Arthur Caplan, Tristam Engelhardt Jr., and James McCartney), (N.Y.: Addison-Wesley, 1981); 545–560. Norman Daniels, *Just Health Care* (Cambridge, U.K.: Cambridge University Press, 1985).

20. Atypical strength or intelligence would, for instance, be considered positive deviations from the norm.

21. Harlan Lane and Michael Grodin, "Ethical Issues in Cochlear Implant Surgery," *Kennedy Institute of Ethics Journal* 7(3) 1997: 231–251, 234.

22. Anita Silvers, " 'Defective' Agents: Equality, Difference and the Tyranny of the Normal," *Journal of Social Philosophy*, 25th Anniversary Special Issue (1994): 154–75.

23. See, for example, Anita Silvers, David Wasserman, and Mary Mahowald, *Disability, Difference, and Discrimination* (Lanham, Md.: Rowman and Littlefield, 1998).

24. For example, it is part of the groundwork for the recent book *From Chance to Choice*, 2000.

25. See, e.g., Haiken, 10; Gilman, *Making the Body Beautiful*; and Barbara Stafford, John La Puma, and David Schiedermayer, "One Face of Beauty, One Picture of Health: The Hidden Aesthetic of Medical Practice," *Journal of Medicine and Philosophy* 14 (1989): 213–30

26. See Diana Dull and Candace West, "Accounting for Cosmetic Surgery: The Accomplishment of Gender," *Social Problems* 38(1) (1991): 54–70, esp. 62.

27. In 2000, nonwhite patients accounted for 13 percent of the national total for cosmetic surgery recipients. The website of the American Society of Plastic and Reconstructive Surgeons reports that 42.4 percent of nonwhites responded that their attitude toward cosmetic surgery was more favorable now than ten years ago (with 40.1 percent holding the same attitude as ten years previously, and only 13.5 percent holding less favorable attitudes). Still, only 11.4 percent of nonwhite respondents said that they have had or think they will have cosmetic surgery at some point in their lives. <www.plasticsurgery.org>. Given the general increase in cosmetic surgery procedures (percent changes between 1992 and 1999 were 389 percent for liposuction, 413 percent for breast augmentation, and 139 percent for eyelid surgeries) and the corresponding change in social norms, however, we might expect the number of nonwhite recipients to increase in the coming years. Haiken's book has a special chapter on what she calls "The Michael Jackson Factor," and she claims that "Jackson is only one among hundreds of thousands of Americans who have attempted, through plastic surgery, to minimize or eradicate physical signs of race or ethnicity that they believe mark them as 'other' (which in this context has always meant 'other' than white)." Haiken, 175–6.

28. Naomi Wolf cites cosmetic surgery brochures that discuss "the Oriental eyelid, the Afro-Caribbean Nose, and the fact that 'the Western nose that requires alteration invariably exhibits some of the characteristics of (nonwhite) noses . . . although the improvement needed is more subtle.'" Wolf, 264.

29. Dawn Perlmutter, "Miss America: Whose Ideal?" in *Beauty Matters*, ed. Peg Brand (Bloomington, Ind.: Indiana University Press, 2000); Paul Taylor, "Malcolm's Conk and Danto's Colors; or, Four Logical Petitions Concerning Race, Beauty, and Aesthetics" in *Beauty Matters* (2000); Stephen J. Gould, *The Mismeasure of Man* (W.W. Norton & Co., 1981).

30. Margaret Little, "Cosmetic Surgery, Suspect Norms, and the Ethics of Complicity," in *Enhancing Human Traits*, ed. Erik Parens (Washington D.C.: Georgetown University Press, 1998), 166.

31. Taylor, 58.

32. Taylor, 58.

33. Taylor, Susan Bordo, *Unbearable Weight: Feminism, Western Culture, and the Body* (Berkeley, Calif.: University of California Press, 1993).

34. Bordo, 253.

35. Bordo, 254.

36. Eugenia Kaw, "Medicalization of Racial Features: Asian American Women and Cosmetic Surgery," *Medical Anthropology Quarterly* 7 (1993): 74–89, 75.

37. Kaw, 75.

38. Kaw, 81.
39. Kaw, 86.
40. Haiken, 92.
41. I say "women" even though this applies as well to men. Although men's use of cosmetic surgery is on the rise, women still account for at least 80 percent of all such surgeries www.plasticsurgery.org⟩.
42. Shlomo Katz and Shlomo Kravetz, "Facial Plastic Surgery for Persons with Down Syndrome: Research Findings and Their Professional and Social Implications," *American Journal on Mental Retardation* 94(2) (1989): 101–10, 103. This review article looks at studies that have measured the effectiveness of such surgeries, both through parental and surgeon subjective report, as well as by changes in appearance ratings by nonrelated controls. In addition to psychological concerns, they also mention that there are physical risks undertaken in the surgeries. Tongue reduction (partial glossectomy) and eyelid corrections both involve the risk of general anaesthesia, and there are other specific risks related to tongue reduction (swelling that obstructs breathing, excessive bleeding, nerve damage, etc.).
43. Katz and Kravetz, 101.
44. Katz and Kravetz, 102. Their review of the literature suggests that while parents will often think their own child sounds better, the majority of nonrelated evaluators (this category also excludes the surgeons who perform the operations) do not find any significant difference in articulation.
45. Despite this negative finding, Katz and Kravetz cite a study by Olbrisch in which he found that 95 percent of the patients' parents said they were satisfied with the operation and would recommend it to others (Katz and Kravetz, 105).
46. Katz and Kravetz, 109.
47. See, for instance, the discussion in Henry Bourguignon, "Mental Retardation: The Reality Behind the Label," *Cambridge Quarterly on Healthcare Ethics* 3 (1994): 179–94, especially, 190.
48. Katz and Kravetz, 107.
49. Barbara Dodd and Judi Leahy, " 'Facial Prejudice' Commentary on Katz and Kravetz," *American Journal on Mental Retardation* 94(2) (1989): 111.
50. Quoted in Judy Shapiro, "New Faces for Down's Kids," *Macleans* 95 (1982): 38.
51. Carol Turkington "Changing the Look of Mental Retardation" *Psychology Today* September: 45–46, 1987, p. 45. This general attitude toward the cosmetic surgeries for Downs children is seconded by Deborah May in her article "Plastic Surgery for Children with Down Syndrome: Normalization or Extremism?" *Mental Retardation* 26(1): 17–19, 1988, p. 19.
52. While this may not be the case currently, given the Katz and Kravetz review article, there is no reason to think improvements could not be made in the effectiveness of the surgeries, just as cosmetic surgery techniques for "normal" adult features have seen significant advances in the last decades.
53. Ronald Strauss "Quasimodo and Medicine: what Role for the Clinician in Treating Deformity?" *Journal of Clinical Ethics* 1(3): 231–235, 1990, p. 233.
54. R.B. Jones "Parental Consent to Cosmetic Facial Surgery in Down's Syndrome" *Journal of Medical Ethics* 26: 101–102, 2000, p. 102.

55. This comparison is not meant to suggest that there are no good reasons for allowing male circumcision while outlawing female circumcision. For instance, although male circumcision is no doubt painful, it does not remove the possibility of sexual pleasure for life, as many forms of female circumcision do, nor does it involve the same degree of on-going health risks as female circumcision (e.g., dysmenorrhea, infertility, incontinence, etc.). For more information, see Sally Sheldon and Stephen Wilkinson "Female Genital Mutilation and Cosmetic Surgery: Regulating Non-Therapeutic Body Modification" *Bioethics* 12(4): 263–285, 1998, especially p. 266.

56. Jones, p. 102.

57. Strauss, p. 233.

58. For discussions of this effect, see Little, 1998 or Silvers, 1998.

59. Or those of their children, in the case of surgery for Down's children. As Frankel and Juengst argue, "Traditionally, parents have been given a good deal of authority in decision-making about functionally unnecessary elective medical treatments for their children. . .[because] parents [have] unique access to information relevant to weighing the risks and benefits involved." The authors argue that some cosmetic surgeries for children may rightfully be seen treatment for the family unit, which is key to the health of the child. Frankel and Juengst, "Cosmetic Surgery for a Fatally Ill Infant" *Journal of Pediatric Ophthalmology & Strabismus* 28(5): 250–254, 1991.

60. Paul Benson, "Feeling Crazy: Self-Worth and the Social Character of Responsibility" in *Relational Autonomy: Feminist Perspectives on Autonomy, Agency, and the Social Self* (eds. Mackenzie and Stoljar), New York: Oxford University Press, 2000.

61. Susan Dodds, "Choice and Control in Feminist Bioethics" in *Relational Autonomy* (eds. Mackenzie and Stoljar), NY: Oxford University Press, 2000.

62. Carol Mcleod and Susan Sherwin, "Relational Autonomy, Self-Trust, and Health Care for Patients Who Are Oppressed" in *Relational Autonomy* (eds. Mackenzie and Stoljar), 2000, p. 260.

63. E.g., Davis, 1995.

64. This is the broad aim of the collection of papers in Mackenzie and Stoljar's book *Relational Autonomy*, 2000.

65. Susan Sherwin, "Normalizing Reproductive Technologies and Autonomy" in *Globalizing Feminist Bioethics* (eds. Rosemarie Tong, Gwen Anderson and Aida Santos), Boulder, CO: Westview Press, 2001, p. 104.

66. According to a website feature at http://www.cbc.ca/programs/ sites/features/ hm_cosmeticsurgery/overview.html, repeat customers currently account for 33% of all cosmetic surgeries. Stafford et al. remark, "The problem with body sculpting is that, like the quest for perfect children, the quest for a perfect body may be unending, since it is based on a fictional norm. Normality actually includes bodily imperfections," 226.

67. Quoted in Haiken, 193.

68. Igor Niechajev and Per-Olle Haraldsson, "Ethnic Profile of Patients Undergoing Aesthetic Rhinoplasty in Stockholm," *Aesthetic Plastic Surgery* 21 (1997): 139–45, 139. They note that "[m]any of our 'foreign' patients speak flawless. . . .

Swedish, seek or keep good jobs, but still feel that their noses are against them," 145.

69. For example, Brand's *Beauty Matters,* Gilman's *Creating Beauty to Cure the Soul,* and *Making the Body Beautiful: A Cultural History of Aesthetic Surgery,* and Nancy Etcoff's *Survival of the Prettiest: The Science of Beauty* (Anchor Books, 2000).

70. I explore this issue in the paper "The Ethics of Making the Body Beautiful: What Cosmetic Genetics Can Learn from Cosmetic Surgery," *Philosophy & Public Policy Quarterly* 21(1) (2001): 21–27.

The Wisconsin Card Sort: An Empirical and Philosophical Analysis of Presuppositions Regarding Flexibility of Cognition

Sara Waller

Introduction

Carnap construed science as fundamentally bound in language, and one of the roles philosophers can play toward bettering science is a careful analysis of definitions and concepts used in the exploration of the empirical world.[1] To that end, I have completed an empirical study in order to examine a few of the concepts fundamental to the production and proclamation of results. While this specific study is well within the boundaries of cognitive neuroscience and uses simple psychometric methods, questions raised here bear on the sciences and social sciences. How do enculturated definitions and assumptions constrain test design and the interpretation of results? How do expectations regarding what is normal or obvious infiltrate and potentially distort our systems of measurement? What are the biases of the experimenters and test designers, how might they be a result of enculturation, and can we ever overcome test bias?

Explanation of the WCST and Some of Its Philosophical Assumptions

The Wisconsin Card Sorting Task (WCST) was developed in 1948 to assess "abstract reasoning ability and the ability to shift cognitive strategies in response to changing environmental contingencies."[2] Today, it is widely used by psychologists for both clinical and research investigations. The test requires an examiner and a single subject seated facing each other at a table. The examiner places four cards on a table for the subject's inspection. Each card shows a different number of shapes, and the shapes on each card are a different color. The subject is then given two decks of cards on which are printed more colored shapes. The subject is asked to match each card in both decks to one of the four key cards. The catch is that no rule is given for matching cards. The subject must decide how to match the cards, and is given feedback from the experimenter: "right" or "wrong." Given the "obvious" choices of matching cards by color, suit, or number, the subject must discern, given feedback, which matching principle to use. So the task measures first the ability of the subject to conclude that there are three possible principles by which to match the cards. But the task is seen as a task of cognitive flexibility. The experimenter switches the conceptual set (color, shape, number) after the subject has completed a set of ten cards according to a certain principle, without warning. So the subject will be happily and correctly matching cards by color, when suddenly the feedback changes from "right" to "wrong." The subject must decide on a new principle for matching the cards. Feedback from the experimenter represents environmental contingency here—there is a change in the success rate of the strategy of the subject, and the subject must switch strategies to reattain success. "Normal" subjects thirteen years old and above easily switch conceptual sets six times during the course of the 128-card task. The usual conclusion is that normal subjects discern the proper conceptual sets easily and can change strategies when the present strategy no longer provides them with the right answers. Many switch sets in fewer than eighty trials, showing quickness in changing from color to shape and shape to number in accordance with feedback.

I shall examine several presuppositions in turn, in order to elucidate philosophical concerns regarding them. First, subjects may or may not recognize color, shape, and number to be the most obvious categories for card matching, and this presence or absence of recognitional format may well be individual, or culture-bound. As Quine[3] suggests, there is a logical possibility that members of different cultures, and indeed individuals within the same culture, may use different principles of individuation in assigning meaning to utterances. What may be a very obvious starting place for one culture may be quite abstract and obscure to another, due to such differences in principles for interpreting or understanding the experienced world. I suggest that by using empirical cues (a strategy not foreign to Quine), we might, at least roughly, define cultural boundaries and from there proceed to guess at what priorities speakers might have in their assignment of meaning and develop useful (though perhaps not truth-indicating) heuristics for determining what individuative principles people might be using. As applied to this study, I divided subjects by ethnicity and socioeconomic status (SES) using standard empirical methods. We can proceed to speculate that, if both factors are highly relevant to performance on the WCST, and that this performance is contingent upon recognizing color, shape, and number as "obvious" starting points, that ethnicity or SES may significantly influence notions of "obvious starting point" and principles of individuation regarding "starting point" altogether.

Another of the main premises of the task is that the subject must look for feedback primarily from the experimenter; that is, the experimenter must be seen as a dramatically important authority regarding relevant environmental cues. Carol Gilligan[4] (in a similar hybrid philosophical/empirical study) documents instances in which people, specifically females, respond to a seemingly boilerplate problem in morality in radically different ways from those expected because different social expectations and mores make girls and women conscious of different aspects of the problem. The problem at hand, the Heinz dilemma, asks subjects to weigh the relative moral value of saving a life against the moral admonition opposed to stealing. Should one steal a drug one cannot afford in order to save one's otherwise terminally ill spouse? Males typically respond with appeals to a universal principle involving the equality of all people and the importance of a single life, as well as the importance of upholding community and moral laws. Females typically respond by rejecting the choice between death and thievery, suggesting compromises and payment plans with the pharmacy and emphasizing the importance of human relationships rather than rules. Gilligan suggests that the difference in responses is a difference stemming from social expectations placed on men and women, claiming that men, as workers in a hierarchically based workplace, will be more likely to obey rules and treat moral problems as equations, while women, as caretakers and homemakers, will be more likely to emphasize relationships and compromise.

Because the WCST is a test that demands acceptance of a hierarchy of feedback, the subject is most likely to succeed if he or she places feedback from the examiner to be very important in determining which card to place in which space. Subjects who resist hierarchy, who are curious and exploratory, and who experiment with alternative forms of feedback (such as the numbers on the back of the cards, other tests taken in the test battery, pictures in the examination room, or the supposed preferences of the examiner) may receive lower test scores than those who accept the hierarchical structure of the experimenter as the sole leader. But members of lower SES groups or ethnic groups historically subject to discrimination may well be predisposed to question authority (as authority figures may be perceived as more likely to be punitive or unjust than helpful) or look for alternative ways to succeed (because experiences with discrimination may have taught them that the most obvious strategies will not work). Such possibilities justify the current study.

The final presupposition of the WCST that I will consider here is this: the task demands that subjects subscribe to a win-stay/lose-shift problem-solving strategy. Roughly, when someone continues to use the same strategy given reward, and shifts strategies given punishment, that person is using a paradigm called "win-stay/lose shift." So, when one has enjoyed the food and service at a particular restaurant, and returns the following week with two friends, win-stay is in use: continue to return to the reward. This strategy is often one we use to evaluate the survival abilities of a person in a real-life situation. For example, we have concern for the cognitive strategy of the battered woman precisely because we see her "lose" time and time again but fail to shift her associations and/or living arrangements.

The second possible problem-solving paradigm—win-shift—seems counterintuitive only at first. Why might a creature change association, environment, or location after a reward? Hummingbirds drink flowers dry—to return to the same flower is not a survival-promoting act. Bees appear to be predisposed to win-shift behavior for similar reasons. Most circumstances involving food or escape are conducive to successful use of the win-shift strategy. One expects a food supply in the wild to be depleted after one visit, and an escape route or strategy to only be surprising to, and thus successful against, a predator only once. While win-stay is conducive to finding a permanent mate, win-shift is propitious to gaining experience with a variety of social encounters.

Of course, other paradigms for solving problems arise according to situations that call for them, and knowing which one to choose also requires cognitive flexibility. Simple problems of memory call for mnemonic devices, problems of metaphysics often call for semantic analysis, and interpersonal problems call for heightened communication and charitable interpretation. The familiar scientific method involves hypothesis formation and data collection. We see that there are many problem-solving paradigms at hand and many strategies that are useful for the survival of the individual and in the survival of the species. The WCST measures only the successful application of win-stay, and so is a measure of only one form of cognitive strategy. These considerations alone raise strong questions as to the adequacy of the WCST to measure such a broad category as "cognitive flexibility." But further, there are reasons why people who have experienced discrimination or have a low SES may not choose win-stay as the best strategy for problem solving. As mentioned, win-stay is a poor strategy to use in a situation in which one feels threatened, such as using an escape route. For someone who does not experience rule-based success in society (e.g. on some days police protect an individual, on others they threaten the same individual), there is no reason to approach the world with a belief in win-stay. Indeed, such an approach to daily life may be considered naïve. Better to stay ahead of whatever threatening events may take place by changing one's strategy often. The possibility that minorities or those from low socioeconomic groups may not subscribe to win-stay in a hierarchical or authority-based situation further justifies the current study.

Thus, the first assumption, that certain aspects of the task are obvious, is surrounded by philosophical questions of indeterminacy and the establishment of a hermeneutic between and across cultures. The second assumption, that the experimenter is a good authority who will provide useful information, is surrounded by (loosely) feminist-style questions of the advisability of acceptance of authority and hierarchy and questions of the oversimplification of our subjects by assuming that they share problem-solving strategies and priorities (such as respect for authority.) The third assumption asks why a test of cognitive flexibility should only measure one form of it; that is, it points out that the definitions we are using to study a certain phenomenon in a scientific way may be too restrictive and thereby produce misleading results.

Background

Familiarity with certain cultures[5] gender[6] and minority status[7] have all been noted as important factors in scores received on various psychometric tasks. SES has been found to be relevant to performance in a number of areas. Using the Hollingshead Four Factor Indicator of Social Status, SES has been found to correlate with high school attrition[8] poor executive function (planning, short term memory, logical, orderly, or causal thinking) in pre-term infants[9] and in predicting schizophrenia.[10] An examination of the WCST in comparison with SES is especially important given these results, as the WCST is considered a definitive measure of cognitive flexibility. Perhaps performance on the WCST reflects SES, and SES, in turn, places restraints on, or encourages profound differences in, problem-solving strategies and lateral (versatile) thinking.

While the WCST is used widely for diagnostic and experimental purposes, and strong correlations have been established between age and number of perseverative errors and between years of education and perseverative errors, little research has been done confirming that it is a test unaffected by race, ethnicity, or SES. The majority of subjects (92 percent) who have taken the test for the purposes of establishing guidelines for the present version of the task have been Caucasian, and while the current manual does an excellent review of scoring norms according to age, education level, and lesion location (as well as several other disorders), there is no report on the performance of subjects relative to socioeconomic status in the current manual (Heaton, et al., 1993). This project, then, lies within the context of Heaton's project of revealing differences in scoring as corresponding with ethnicity and SES.

Subjects

Sixty-nine subjects ranged in age from 15 to 69 (mean age 25.55, standard deviation 9.33 years, median 22 years, mode 18 years). Thirty-five of all subjects were in the 18–25-year age range (50.5 percent). Twenty-three subjects were under age 20, twenty-four subjects were 20–29, sixteen subjects were 30–39, five subjects were 40–49, and one subject was 60. Subjects were grouped by self-identified race. Thirty subjects identified themselves as African American, eighteen subjects identified themselves as Hispanic, eleven identified themselves as Caucasian, and nine identified themselves as Asian. Twenty subjects were ranked in the first category (High) in SES, twenty-two were ranked in the second, twelve were ranked in the third, eleven were ranked in the fourth and two were ranked in the fifth (Low).

Method

Subjects were asked to fill out a modified version of the Hollingshead Social Status form, and take five subtests (Verbal, Bead Memory, Quantitative, Memory for Sentences, and Pattern Analysis) of the Stanford Binet Intelligence Scale (Fourth Edition, 1986) along with the Wisconsin Card Sorting Task (1993). Upon completion of the WCST, subjects were asked to write down what they were thinking as they sorted the cards and identify any principles of sorting they were using to sort the cards correctly. Subjects were asked to rank categories of the task in order of importance and whether the examiner was seen as attempting to "trick" the examinee during the WCST. Informed consent was obtained for all participants prior to participation in the study. The first sixty subjects were reimbursed $10 for their time and efforts, and the remaining subjects were given extra credit in a college course. Subjects were informed of their IQ scores either verbally or by confidential mailing, or both, unless they indicated during the test session that they did not want to be informed of their final scores. Subjects were screened for color blindness. One subject who was color blind but successfully identified red and green cards on the WCST and red, white, and blue beads on the Stanford Binet bead memory section was allowed to proceed. This subject passed the WCST and scored within normal range on bead memory and was included in the final computations.

Results

Significant correlations at the .05 level were found between IQ and Ethnicity, IQ and SES, Passing the WCST and SES, WCST Trials to Completion and SES, IQ and Passing the WCST, IQ and WCST Trials to Completion. (Table 1) An ANOVA was done for SES and WCST Trials to Completion with a significant result (.001). Then a contrast analysis was performed for performance for SES of levels one and two as opposed to three, four and five. The results were significant at the .001 level. (Table 2)

A second ANOVA comparing ethnicity and trials to completion was performed. (Table 3) The results here were not significant (.077), but trended toward significance. Because results did trend toward significance, one further exploratory contrast analysis was performed, with equal variance assumed. The Caucasian group (group three) performance means were compared to means for African Americans (group one), Hispanics (group two) and Asians (group four), and a significant difference was found between group three and the other groups.

A regression analysis placing SES as the first variable and Ethnicity and IQ as the remaining variables showed that SES had the most significant effect on trials to completion. (Table 4)

Premises of the WCST include use of the win-stay strategy, including feedback from the examiner as very important information as to how to match the cards, and identifying color shape and number as the most obvious or prominent categories by which to sort the cards. How was the explicit adoption of these premises relevant to performance on the WCST? (100 percent) of the subjects noted the importance of feedback in the essay section of the questionnaire ("Describe your thinking as you sorted the cards"), including subjects who did not pass. This suggests that subjects did

TABLE 1 Correlations

N=69 in all cases. Significant correlations are marked with an asterisk.

	Ethnicity	SES	IQ	Pass WCST	Trials to Completion
Ethnicity Pearson Correlation		−.85	.467*	.163	−.226
Ethnicity Sig (2 tailed)		.486	.000	.181	.061
SES Pearson Correlation	−.085		−.267*	−.331*	.408*
SES Sig (2 tailed)	.486		.026	.005	.000
IQ Pearson Correlation	.467*	−.267*		.275*	−.292*
IQ Sig (2 tailed)	.000	.026		.022	.015
Pass WCST Pearson Correlation	.163	−.331*	.275*		−.814*
Sig (2 Tailed)	.181	.005	.022		.000
Trials to Completion Pearson Correlation	−.226	.408*	−.292*	−.814*	
Sig (2 Tailed)	.061	.000	.015	.000	

not necessarily have a fundamentally different assessment of how to start the task or that feedback was important in completing the task. None of the subjects reported feeling suspicious of the examiner or feeling that the examiner was trying to trick them. Thus, the fact that SES was the single best predictor of test performance does not lead us to a notion of distrust of authority figures as an explanation for task failure. However, reactions of distrust are probably not completely measured by self-report, as the subject may not be aware of mistrust or be hesitant to report distrust of the examiner to the examiner.

Because SES was the best predictor, we can certainly ask questions regarding the use of a different strategy, such as win-shift, being favored by those who do not have high status in American society. Subjects who failed

TABLE 2 SES and Trials to Completion ANOVA

	Sum of Squares	df	Mean Square	F	Sig
Within	9945.433	4	2486.358	5.430	.001
Between	29303.118	64	457.861		
Total	39248.551	68			
Contrasts Test	Value	Std. Error	t	df	Sig (2 tail)
Assume Equal Variance	14.68	6.64	2.211	64	.031
Do not Assume Equal Variance	14.68	4.95	2.966	13.524	.011

would often comment that the game did not make sense, or that there were no rules, and express frustration. Subjects who passed easily were very confident that there were rules, that the game could be mastered, and that they simply needed to try various ideas until they hit on the right one. This could suggest a certain amount of learned helplessness manifest among those of low SES—they tended to believe that the task was fundamentally irrational, that they would be told that they had made the wrong choice again and again for 128 trials with no hope for correction or redemption. Much like a Kafka character, or an example of Camusian or Sartrean angst at being confronted with an absurd world, they placed card after card hopelessly. But this is not so unlike many of their life experiences, in which they may be treated irrationally by power figures or "rule systems" that are set up for them to fail. Such subjects may already have an expectation that there is no strategy, neither win-stay nor win-shift, that will allow them to profit, because all systems are either irrational or biased against them or both. Thus, being asked to follow a hierarchical or rule-based system may

TABLE 3 Ethnicity and Trials to Completion ANOVA

	Sum of Squares	df	Mean Square	F	Sig
Within	3887.873	3	1295.958	2.382	.077
Between	35360.678	65	544.010		
Total	39248.551	68			
Contrasts Test	Value	Std. Error	t	df	Sig (2 tailed)
Assume Equal Variance	−16.38	7.79	−2.103	65	.039

TABLE 4 Regression

Dependent variable is number of trials to completion of the WCST

	Model	Sum of Squares	Df	Mean Square	F	Sig
SES	Regression	6547.932	1	6547.932	13.416	.000
SES	Residual	32700.619	67	488.069		
SES	Total	39248.551	68			
SES, ETH	Regression	8000.379	2	4000.190	8.449	.001
SES, ETH	Residual	31248.172	66	473.457		
SES, ETH	Total	39248.551	68			
SES, ETH. IQ	Regression	8509.283	3	2836.428	5.998	.001
SES, ETH. IQ	Residual	30739.268	65	472.912		
SES, ETH. IQ	Total	39248.551	68			

still be an important factor in the intimidation of or triggering of hopelessness of subjects with low SES. Further, low SES subjects may reject the premise that there is any strategy that will be helpful; thus, the WCST may be measuring a learned helplessness rather than cognitive flexibility, though these two notions may be related.

In a follow-up question, subjects were asked to rank factors in the test in order of importance. Nearly all, including those who failed or had an extremely high number of trials to completion, ranked color, shape, and number in some order. Of the seventeen subjects who did not pass the WCST, six (35 percent) added an additional matching principle, and twelve (75 percent) reported matching cards by only one or two of the three matching principles. This provides support for our first cluster of philosophical questions surrounding the task—that some subjects did not recognize color, or shape, or number as important in sorting the cards. This suggests both that the task demands certain building blocks to be in place that may be irrelevant to cognitive flexibility and that these building blocks may not be ready to hand for those in lower SES groups. This leads to the alternative explanation that those who score poorly on the WCST are using slightly different principles of individuation for the cards (though these principles, unlike Quine's, may ultimately be empirically detectable) and that the test is not one of cognitive flexibility but of card classification or individuative system. Once the cards have been individuated differently, the cognitive strategy used to classify the cards may well be completely indiscernible; that is, the subject may actually be using an outstanding version of win-stay, but the experimenter will never be able to detect this, and the subject will fail because her ontology is in slight variation with the expected ontology.

Closing Experimental Remarks

Given the small sample sizes of some of the socioeconomic and ethnic groups included in this study, further research is certainly needed to confirm findings for SES and ethnicity. Also, a refined post-WCST questionnaire would bring to the fore exactly how subjects evaluated feedback from the examiner and would provide further evidence for or against the thesis that SES and/or ethnicity has an effect on how feedback from an examiner is perceived and utilized.

Acknowledgments

Special thanks to Dr. Amy Schatz and Dr. Angela Ballantine, and the UCSD Pediatric Neurology Research Group, for providing interrater reliability and other positive suggestions for this study. Thanks to Robert Heaton for helpful comments. Any errors are mine and not theirs.

Notes

1. Rudolph Carnap, "The Elimination of Metaphysics," in *Logical Positivism*, ed. A.J. Ayer (New York: Free Press 1959), 77.
2. Robert Heaton, et al., *Wisconsin Card Sorting Test Manual* (New York: Psychological Assessment Resources, Inc., 1993), 1.
3. W.V.O. Quine, *Ontological Relativity and Other Essays* (New York: Columbia University Press, 1969), 34.
4. Carol Gilligan, *In a Different Voice* (Cambridge, Mass.: Harvard University Press, 1982).
5. Ronald Samuda, Reuven Feuerstein, Alan Kaufman, John Lewis, and Ron Sternberg, *Advances in Cross-Cultural Assessment* (New York: Sage Publications, 1998).
6. David Sadker, *Failing at Fairness* (New York: Macmillan Press, 1994).
7. Stephen J. Gould, *The Mismeasure of Man* (New York: W.W. Norton and Co., 1996).
8. Amy Elizabeth Plog, The Study of Attrition in a Preventative Intervention for Boys at Risk for Conduct Disorder (Ph.D. diss., University of Houston 1996).
9. Brenda M.Wall, Executive Function of Children Born Preterm. (Ph.D. diss., Fordham University 1996).
10. Deana S. Benishay, Biosocial Antecedents of Schizophrenia-Spectrum Personality Disorders: A Longitudinal Study (Ph.D. diss., 1997).

Additional Sources

Block, N.J., and Gerald Dworkin. *The IQ Controversy*. New York: Random House, 1976.

Butler-Omolulu, Cynthia, Joseph Doster, and Benjamin Lahey. "Some Implications for Intelligence Test Construction and Administration with Children of Different Racial Groups." *Journal of Black Psychology* 10 (2) (1984): 63–75.

Chase, Allan. *The Legacy of Malthus: The Social Costs of the New Scientific Revolution*. New York: Alfred A. Knopf, 1977.

Franklin, Godfrey, and Mamie Hixon. "Your Dialect Could Place You on the Wrong Side of the Intelligence Bell Curve." *Negro Educational Review* 50 (3–4) (1999): 89–101.

Goodland, John, and Pamela Keating. *Access to Knowledge: The Continuing Agenda for Our Nation's Schools*. New York: College Entrance Examination Board, 1994.

Jensen, Arthur R. *Bias in Mental Testing*. New York: MacMillan Pub., 1980. Especially, 732.

Martens, Brian, Emily Steele, Doreen Massie, and Maureen Diskin. "Curriculum Bias in Standardized Tests of Reading Decoding." *Journal of School Psychology* 33 (4) (1995): 287–96.

Scales, Alice M. "Alternatives to Standardized Tests in Reading Education: Cognitive Styles and Informal Measures." *Negro Educational Review* 59 (2–3) (1987): 99–106.

Spearman, Charles Edward. *The Abilities of Man; Their Nature and Measurement*. New York: Macmillan Co., 1927.

Trout, J.D. "Diverse Tests on an Independent World." *Studies in the History and Philosophy of Science* 26 (3) (1995):407–29.

Weiss, John G. "It's Time to Examine the Examiners." *The Negro Educational Review* 58 (2–3) (1987):107–24.

Geography and Ideas of Race
Naomi Zack

Race Factors

Ever since the early nineteenth century, when the idea of race as biological human difference became effective for the organization of society, the public has been confident that evident human differences in skin color and morphology could be explained by descriptions of general race factors knowable to scientists. For example, it is believed that black people have some physical thing that makes them black and causes dark skin shades. General race factors, such as blackness, are believed to determine racial membership and cause specific racial traits.

The public has never explicitly required that general race factors be observable to it, but is usually satisfied if the specific race traits are observable; for instance, it is supposed that anyone can see that a black person has "dark" skin. The public will even accept racial identities at odds with observable physical traits; for instance, fair-skinned blacks and dark-skinned whites. Still, in the jumble of everyday life, average people sort others into races, based on their appearances, without questioning the existence of general race factors, or separating the assumption of their existence from the process of racial sorting. The public does not have theories about its epistemology or ontology concerning what it thinks is race, and it doesn't seem to need them, either. In ordinary experience, the biological taxonomy of race appears simply to be given in hereditary human differences that are evident "on the street." Anyone who denies the reality of race is taken to be denying common evidence of biological variety, which is absurd to do. Not surprisingly, the main business of race concerns the different attitudes and behavior that are attached to differences in appearance that are assumed to be racial. This is the primary sense in which "race matters" in daily life. That human biological variation in itself does not prove the existence of race is a fact that barely registers. A denial of the existence of biological variation would entail a denial of the existence of race. However, denial of the existence of race does not entail denial of the existence of human biological variation.[1]

The epistemology of race is the process of sorting people into different races and the criteria for racial membership or identity. The ontology of race rests on evidence for the existence of two or more races—evidence for racial taxonomy in the first place—and its basic premise is either; "Yes there are human races," or, "No, there are not human races." Ashley Montagu and other mid-twentieth century anthropologists pointed out that any racial taxonomy would require factual support before it was constructed *as a scientific taxonomy*, and the factual support would have to be of a scientific nature and not just ordinary beliefs that there are human races. Such support did not exist when Montagu wrote, and it is still not in evidence, although many continue to believe that it is.[2] It is therefore important to understand precisely what facts are lacking and where the conceptual gaps occur in the thought of those who believe that there is now scientific evidence for the human racial taxonomy.

Scientists usually prefer to posit observed, rather than unobserved, explanatory entities But, if there is evidence that a cause, X, exists, a present technological inability to observe X need not mar its empirical respectability. Viruses and genes are examples of explanatory entities that were accepted before they could be directly observed. Atoms and subatomic particles may never be directly observed and may in principle be impossible to ever observe. Therefore, if there were evidence for the existence of general race factors, the fact that they had not, or could not, be directly observed would not necessarily impugn their scientific status. If there were no evidence for general race factors, statistically reliable specific factors might serve as a scientific foundation for the human group differences that the public believes are racial—if there were independent evidence for their existence. Statistically reliable specific race factors would be a weaker scientific foundation for common sense racial taxonomy than general race factors, but they might be useful. For instance, there might be specific biological traits that were present only in all, or even most, people who were socially identified as black, and descriptions of those traits could be used as a biological foundation for definitions of blackness. Knowledge of specific racial factors would be useful beyond definitive purposes if they were known to be connected with other human traits distinctive to socially iden-tified racial groups. Generally speaking, this kind of connection is the min-imum of what the educated public often assumes, is, or someday will be, the result of scientific inquiries about race.

Leaving aside racial essences, which could not be empirical entities, there have been four bases for ideas of physical race in common sense: geo-graphical origins of ancestors; phenotypes or physical appearance of indi-viduals; hereditary traits of individuals; genealogy. Ever since race became a subject in human biology and anthropology, the scientific search for gen-eral, as well as statistically reliable specific race factors, has focused on these four bases in varied combinations. Eighteenth and nineteenth century white supremacist scientists claimed that Europeans were culturally and intellec-tually superior to Africans and Asians, as well as healthier and more beauti-ful physically.[3] During the nineteenth century, most scientists assumed that human races existed and that each race was distinguished by hereditary fac-tors that determined physical appearance, as well as culture and psychology. Quantitative attempts to measure physical racial differences proceeded from assumptions that scientists could distinguish among members of the differ-ent races, based on their appearance and geographical origins, before they made their measurements. As Stephen Jay Gould and others have shown, these assumptions of prescientific difference privileged whites before the anthropometry of cranial capacity (believed to indicate brain size, which in-dicated intelligence) and limb proportion was undertaken. The actual mea-surements were then interpreted in ways that supported presumptions of white supremacy, and the raw data was also recorded inaccurately, and was at times falsified, to corroborate the presumptions.[4]

In the early twentieth century, cross-cultural anthropological studies made it evident that cultural and psychological racial distinctions were the result of historical events and contingencies only, so that they could not be inherited along with the physical traits that were still taken to indicate racial identity.[5] This put the scientific burden of biological racial difference on physical biology alone. Developments in neo-Mendelian heredity the-ory, human genetics, and evolutionary theory opened the question of whether there were human races in even a restricted physical biological sense. However, assumptions that geographical ancestral origin is an em-pirical basis for racial difference and identity continue to be held today by humanistic and scientific theorists of race and by the lay public. In this paper, I will examine the geographical basis of ideas of race, in history and contemporary science.

The History of Geography as a Basis of Race

Most contemporary theorists of race, and critics of racialism and racism, recognize that the modern Western idea of race was constructed, at least in part, as a justification for the unjust treatment of victims of slavery, colonialism, and genocide in an age when Enlightenment principles required universal justice and equality. The seventeenth-century rights of political subjects, the eighteenth-century rights of men, and the nineteenth- and early-twentieth-century democratic notions of individual rights were meant to be restricted to white males, preferably property owners. The rationalization of oppression and unequal treatment of women and nonwhites, compared to white males, has come to be understood as an application of crudely drawn abstract biological taxonomies that favor white males. (Indeed the liberatory scholarship and activism of the last third of the twentieth century developed from that historical thesis.) But within the liberatory tradition, sometimes insufficient attention has been paid to the original and ongoing material conditions of oppression. Geography, as a fact in the world that was a parameter of original oppression and as an idea later connected to falsely essentialist abstract biological taxonomies, is a crucial material condition in this neglected sense.

Nonwhite racial categories were imposed on people who lived in, or came from, particular places. Those places—Africa, Asia, America—and the conditions of their discovery by Europeans were conceptualized before biological racial categories were constructed, so that ideas about the places grounded the later taxonomies of biological races. The first modern European attempts to classify and describe non-European groups were made by travelers and traders during the European "Age of Discovery." Notions of place preceded even prototypical notions of race. We can assimilate the tones of shock, wonder, and titillation in descriptions of newly discovered non-European places to later racisms that accompanied later ideas of race, but it is anachronistic to do so. Consider Amerigo Vespucci's 1503 *Mundus Novis* (Letter on the New World), which so captured the imagination of Europeans that the continent that would otherwise have been named for Columbus was called "America." Amerigo described human inhabitants of that continent, not as instances of an abstract classificatory category, but as the denizens of the geographical place that was his main interest:

> Part of this new continent lies in the torrid zone beyond the equator toward the Antarctic pole, for it begins eight degrees beyond the equator. We sailed along this coast until we passed the tropic of Capricorn and found the Antarctic pole fifty degrees higher than the horizon. We ad-

vanced to within seventeen and a half degrees of the Antarctic circle, and what I there have seen and learned concerning the nature of those races, their manners, their tractability and the fertility of the soil, the salubrity of the climate, the position of the heavenly bodies in the sky, and especially concerning the fixed stars of the eighth sphere, never seen or studied by our ancestors, these things I shall relate in order.[6]

By 1748, when Georges-Louis Leclerc, Comte de Buffon, wrote *A Natural History, General and Particular,* he was able to catalogue differences in races to correspond with differences in geography, so as to suggest a research program. Buffon's use of geographical differences to explain racial differences was undertaken as a defense of his monogenicism, and that is how it is usually interpreted. But his assumed close causal connection between geography and race deserves attention in its own right because it was to become so widespread. Buffon believed not only that continental geographic differences accounted for differences among the major races, but he attempted to relate differences within continents to differences within individual races:

> Africa is not less singular for the uniformity in the figure and colour of its inhabitants, than Africa is remarkable for the variety of men it contains. This part of the world is very ancient, and very populous. The climate is extremely hot; and yet the temperature of the air differs widely in different nations. Their manners also are not less various, as appears from the description I have given of them. All these causes have concurred in producing a greater variety of men in this quarter of the globe than in any other.[7]

After Buffon, Johann Friedrich Blumenbach began his version of monogenicism with the speculation that the primary human race was white. Blumenbach then accounted for the existence of nonwhites with a theory of the ways in which white subgroups had *degenerated,* due to different geographical conditions. Like Buffon, Blumenbach catalogued races on a geographical basis, but he speculated that geographical factors were the causes of specific racial degenerations. Blumenbach's first two maxims, or "corollaries" to his theory of degeneration, make this clear: In the first maxim, he posits geographical variations over time as causes of complex human development, in general:

> (1) The more causes of degeneration which act in conjunction and the longer they act upon the same species of animals, the more palpably that species may fall off from its primeval conformation. Now no animal can be compared to man in this respect, for he is omnivorous, and dwells in every climate, and is far more domesticated and far more advanced from his first beginnings than any other animal; and so on him the united force of climate, diet, and mode of life must have acted for a very long time.

Blumenbach's second maxim posits geographical variation as a cause of variations within geographically defined human groups, or races.

> (2) On the other hand an otherwise sufficiently powerful cause of degeneration may be changed and debilitated by the accession of other conditions, especially if they are as it were opposed to it. Hence everywhere in various regions of the terraqueous globe, even those which lie in the same geographical latitude, still a very different temperature of the air and equally different and generally a contrary effect on the condition of animals may be observed, according as they differ in the circumstances of a higher or lower position, proximity to the sea, or marshes or mountains, or woods, or of a cloudy or serene sky, or some peculiar character of soil, or other circumstances of that kind.[8]

The important difference between Buffon, writing in 1748, and Blumenbach, writing in 1776, is that Blumenbach, with his notion of degeneration, attached a scale of human worth to geographical difference. By the time we get to Hegel, writing in the 1820s, the influence of geography in the formation of races is so great that the full flower of civilization could have been possible only in Europe. Hegel takes European superiority for granted and looks for a geographical explanation in comparing Asians with them.

> Since no one particular type of environment predominates in Europe as it does in the other continents, man too is more universal in character. Those particular ways of life which are tied to different physical contexts do not assume such distinct and peculiar forms as they do in Asia, on whose history they have had so great an effect; for the geographical differences within Europe are not sharply defined. Natural life is also the realm of contingency, however, and only in its universal attributes does it exercise a determining influence commensurate with the principle of the spirit. The character of the Greek spirit, for example, grew out of the soil of Greece, a coastal territory which encourages individual autonomy. Similarly, the Roman Empire could not have arisen in the heart of the continent. Man can exist in all climates; but the climates are of a limited character, so that the power they exercise is the external counterpart to man's inner nature. Consequently, European man also appears naturally freer than the inhabitants of other continents, because no one natural principle is dominant in Europe. Those distinct ways of life which appear in Asia in a state of mutual conflict appear in Europe rather as separate social classes within the concrete state. . . . The sea provides that wholly peculiar outlet which Asiatic life lacks, the outlet which enables life to step beyond itself. It is this which has invested European political life with the principle of individual freedom.[9]

Thus, for Hegel, geography is not merely a cause of physical difference as it was for Buffon, or of physical and moral difference, as it was for Blumenbach, but, in the (convenient) language of other parts of his philosophy, an

expression of *spirit*, of which, in the case of race, physical and moral difference is the coexpression. Also, for Hegel, Asia is the only contender to Europe, because he has already removed Africa from the self-aware progres-
sion of human history, again based on geography (as that coexpression,
with physical and moral difference, of spirit):

> *Africa Proper* is the characteristic part of the whole continent as such. We
> have chosen to examine this continent first, because it can well be taken
> as antecedent to our main enquiry. It has no historical interest of its own,
> for we find its inhabitants living in barbarism and slavery in a land which
> has not furnished them with any integral ingredient of culture. From the
> earliest historical times, Africa has remained cut off from all contacts
> with the rest of the word; it is the land of gold, forever pressing in upon
> itself, and the land of childhood, removed from the light of self-conscious
> history and wrapped in the dark mantle of night. Its isolation is not just a
> result of its tropical nature, but an essential consequence of its geograph-
> ical character.[10]

Geography as a Basis of Race in Contemporary Science

Early scientific accounts of the geographical history of *Homo sapiens* were
connected with core hypotheses about whether apparent racial differences
are the evolutionary effect of migrations from one original location, or the
result of different geographical origins for different races. In the eighteenth
and nineteenth centuries, *polygenicists* argued that distinct human racial
groups had different ancestral geographical origins, and *monogenicists* argued that all human groups derived from the same ancestral geographical
origin.[11]

Intuitively, it would seem that polygenicism, in positing independent
origins for human races, would be more supportive of racism than monogenicism, which claims one origin. Richard Popkin claims that polygenicism, as well as contemporary theories of racially inherited differences in
intelligence, need not result in racist doctrines. Popkin cites Frederick
Douglass as a prior authority for the general view that strong differences
between whites and nonwhites do not abrogate the moral status of nonwhites.[12] While this position of Popkin, and before him, Douglass, is logically coherent and motivated by principles of human justice, its primary
concern is racism, or egalitarianism, *on a foundation of settled categories of
race*. However, when those settled categories themselves are in question,
logically and empirically, it is important whether *Homo sapiens* originated
in one place, or in several geographic locations that correspond to the ancestral origins of groups identified as races in society. In itself, confirma-

tion of polygenicism would have been more supportive of human racial taxonomy, than confirmation of monogenicism. However, polygenicism, or the theory of distinct human group origins, has been discredited in biological anthropology since about 1960.[13]

The present consensus in biological anthropology is that all human groups originated from common ancestors in Africa. However, there is considerable disagreement about the length of time that groups, presently on continents other than Africa, occupied those continents. Roughly speaking, out-of-Africa anthropologists believe that modern humans originated in Africa about 100,000 years ago, and began to migrate to other continents about 70,000 years ago, while multiregionalist anthropologists argue that after an original migration from Africa, groups evolved on different continents over a period of one to two hundred thousand years.

According to the contemporary multiregionalist hypothesis, European, Asian, and African Homo sapiens populations that evolved on different continents, after the original African migration, evolved as one species, because there was continual gene flow among groups over the history of their evolution.[14] In contrast, the out-of-Africa theorists maintain that Homo sapiens probably originated about 500,000 years ago, and that all modern humans originated in Africa between 140,000 and 100,000 years ago. In this view, migration from Africa is held to have occurred at the following approximate dates: 70,000 years ago, humans migrated from Africa to Asia; 43,000 years ago, humans migrated from Asia to Europe; 15–50,000 years ago, humans migrated from Asia to the Americas.[15] These dates are approximate because recorded human history goes back only 10,000 years, and both the archeological and biological evidence is incomplete. The out-of-Africa hypothesis is partly based on the unlikelihood of the independent development of the genetic similarities which now exist among African, Asian, European, and American Indian groups.[16] Still, multiregionalist theorists claim that genetic similarities among groups can be accounted for on the basis of gene flow, the out-marriage or breeding between members of different continental groups. Multiregionalists calculate that this gene flow need to have occurred relatively infrequently, even only once in a generation, to produce the genetic similarities necessary to consider all groups as members as the same evolving species.[17]

Luigi Cavalli-Sforza, an out-of-Africa theorist, explains that the longer the period of time a population has occupied a continent, according to archeological data, the greater the genetic distance, in terms of blood groups and protein polymorphisms, of that population from present occupants of the continent from which it migrated. Thus, blood and protein genetic distance is proportional to the time ancestrally related populations have been reproductively separated from each other. Cavalli-Sforza notes that most nongenetic, or phenotypic, data, such as measurements of height and other anthropometric traits, including what are commonly taken to be racial traits, are a less reliable gauge of differences between populations than genetic data, because the anthropometric traits have been influenced by environmental effects on individual development, and by genetic modification due to natural selection (adaptation).[18] In contrast, multiregionalists point out that Cavalli-Sforza's equation of genetic distance with the time of reproductive separation between populations is based on an assumption that human continental groups have been reproductively isolated. With gene flow among groups, the relationship of current geographical populations to older populations in the same area need not be a matter of direct descent. Some writers claim that greater genetic distances between non-African and African populations are not the effects of when migrations occurred (which is how the out-of-African theorists interpret the data) but the effects of a larger African population and greater gene flow out of Africa than into Africa.[19]

While natural selection changes the genetic makeup of populations, that kind of genetic change is not useful for determining the ancestral origins of populations. This consideration is highly relevant to attempts to ground racial identity on the geographical location of ancestors, because racial criteria are phenotypical and can be measured only "anthropometrically." It also implies that ancestrally close groups might display apparent racial phenotypic differences if present members are descended from populations that had to adapt to very different environments.[20]

To the lay person, the fact that what are taken to be physical racial traits are not themselves good indicators of ancestral lineage is paradoxical, especially since the genes for such traits could be identified. But many of those socially designated racial traits are present in soft tissue, which cannot be easily reconstructed from osteological (skeletal) remains. Moreover, a major part of the scientific study of human evolutionary history has been to determine the ancestry of what are taken to be racial groups and to make the determination in ways that will confirm or disconfirm social racial taxonomy. And, it is crucial in studies intended to confirm or falsify the hypothesis that there are human races, that once race has been identified in the ordinary way, other biological traits present in racially identified groups be used to track their lineage over time and distance. This difference in identifying and tracking traits protects the race-confirming or race-denying conclusion from being circular; it ensures that an assumed taxonomy is not being projected onto those facts that are supposed to be the foundation for the taxonomy. There are devastating epistemological problems with the race-identifying or phenotypical traits, although that is not the point at issue, here.[21] Here, it is important to keep in mind that the geographical study of human evolutionary history is not the same thing as confirmation or falsification of the present existence of race. Distinct social races could have geographically distinct ancestry, but that doesn't provide a scientific basis for race, unless there is something scientifically racial about the ancestry.

To return to Cavalli-Sforza's account, the best genetic materials for tracking population lineage are hereditary mutations in genetic material, which have no adaptive function because they have no protein-producing capability. An apparent exception to this rule is the use by evolutionary theorists of HLA (human leukocyte antigens) gene groups, which are important in human immunology. But, HLAs change as much, or more, due to chance as natural selection, and it is their frequent, nonfunctional mutations that make them suitable for tracking populations.[22]

Gene frequencies for ABO and RH blood groups, and genes that code for proteins and enzymes, are usually measured within populations. DNA has more recently been used to measure genetic distances between individuals, by counting the different mutations in DNA sequences. (DNA is made up of ACG and T nucleotides in different sequences. A complete set of human chromosomes contains three billion nucleotides, the human genome.)[23] The Human Genome Diversity Project, begun in the early 1990s, compared world populations that corresponded to social racial groups for mutational genetic difference. As Cavalli-Sforza relates, all blood protein and DNA marker comparisons result in world trees, within which the greatest genetic difference exists between present occupants of Africa and non-Africans. These measures of genetic difference are usually taken to confirm the out-of-Africa thesis about human origins.[24] But multiregionalists argue that the greater genetic distance between Africa and all other populations does not necessarily indicate when population splits occurred because it could be due to a larger African population with greater genetic diversity and more gene flow out of it than into it.[25]

The analyses of hereditary blood proteins were used to identify and compare human lineages, both within and between continents, beginning in the 1960s. But, in the past two decades, it has become technologically possible to track descent by the analysis of mitochondrial(mt) DNA. Mt DNA does not break up and recombine in the formation of reproductive cells, as most other genetic material does, but is instead directly passed on by mothers to children. Since women with different mtDNA have different mothers, a mitochondrial lineage is a direct maternal lineage. MtDNA is subject to a high rate of mutation and is present in each human cell. The date of "Mitochondrial Eve," or "African Eve," the common ancestress of present humans, which Cavalli-Sforza accepts as 143,000 years ago, was arrived at by counting the number of mutations separating Africans from non-Africans and comparing it with the number separating chimps from humans, in calibration with archeological dating. All females have two X chromsomes, one from each parent. Males have XY, an X from their mothers and a Y from their fathers, so that the Y chromosomes in males can come only from their fathers. Based on a single nucleotide mutation of the Y chromosome, there is also believed to be "African Adam," the common ancestor of present male humans, who lived about 144,000 years ago (and was not a mate of Eve).[26]

Further findings suggest that there were three African Eves, one of whose descendants branched into six Asian Eves, of which one descendant branched into nine European Eves. Several Asian Eve lineages continued directly into the Americas, with no further mtDNA mutations to mark branching. There were three African Adams, one of whose descendants branched into seven Asian Adams, whose lineage continued into Oceania, Europe, and the Americas.[27]

In the out-of-Africa hypothesis, mitochondrial African Eve would be the ancestress of all present humans, both male and female. But since only males have Y chromosomes, chromosomal African Adam, while the ancestor of all present males, may not have been the ancestor of all present females. (That is, African Adam and his descendants may have mated with the daughters of another Y line that has since passed out of existence.)

The absence of distinct mutations for European male lineages and for both American male and American female lineages is difficult to interpret, beyond the conclusion, based on mtDNA and Y nucleotide mutations, that they have no branches within them. There may or may not have been other genetic changes, not accompanied by identifiable chromosomal mutations, after the time of population migration from Asia to Europe and the Americas. But, as well, there may or may not have been other genetic changes that accompanied the identified genetic mutations for both males and females during the times marked by distinct mtDNA and Y nucleotide chromosomal lineages.

Finally, relying once more on Cavalli-Sforza, there are microsatellites, short repetitive sequences of DNA, which contain errors or mutations that can serve as genetic markers, because once errors are made in the copying of DNA sequences, during the production of reproductive cells, the errors become a permanent part of DNA sequences. Such errors are the basis for absolute genetic dating, if a rate of mutation can be ascertained. Absolute dating of early humans does not depend on calibration with archeological evidence, although it places the origins of modern humans in Africa 80,000 years ago, a date that does accord with the independent archeological data.[28]

Again, multiregionalists dispute that Cavali-Sforza's and related studies of mitrochonrial Eve support a recent African origin of modern *Homo sapiens*, which then populated the rest of the world.[29] The length and details of modern *Homo sapiens* history are empirical questions that may be impossible to settle. Still, all contemporary biological anthropologists posit African origins, at some point in the past. But, and this is the important point, there is no evidence that African *Homo sapiens* origins are origins of a particular race. None of the genetic data used to reconstruct human natural history relies on racial classifications as they are commonly understood, and neither has this data been shown to have any connection with such classifications. Human origins are by definition geographical, and race has been metaphorically associated with geography, but that is far from a causal connection. The out-of-African theorists track human migrations on the assumption that populations can be studied as self-contained breeding units, but nothing in their data or its interpretations supports social racial taxonomies. The multiregionalists reconstruct a longer human history on the presumption of gene flow among populations. They are more likely to make use of antropometric data in identifying and comparing present geographic populations, but they have less reason to believe that such phenotypic traits have remained the same over the history of a population in a given geographic areas, precisely because of the gene flow among populations that they posit.[30]

The Continuing Lure and Lore of Geography

What does the scientific view of human geographical migration and genetic mutation indicate about the existence of race? Commonsense racial categories have a history originating in European colonization of Africa and the Americas and cultural distance from (if less successful domination of) the Middle East, China, and Japan. Those in the West have divided human groups into biological races, partly on the basis of continental geography, even though the reasons for this division were originally economic, political, and religious. Biological ideas of race were developed in the eighteenth and nineteenth centuries, well after the periods of exploration and colonization. (And, of course, the process of racial identification was not culturally symmetrical between Europe and its subalterns. The non-European groups who came to have black, Asian, and Indian racial identities imposed on them did not originally consent to those designations, although they developed ideologies of resistance and liberation within intellectual traditions based on reinterpretations of their imposed nonwhite racial identities.[31]) Nonetheless, the primary cut between Africans and non-Africans in genetic distance and the attendant genealogical branching, with smaller amounts of genetic difference among populations in Asia, Europe, and the Americas, at first glance seems to support historical and commonsense geographical racial taxonomy. But, how exactly?

The consensus that *Homo sapiens* originated in Africa has prompted some writers to claim that the sole human race is the Negro or black one. Lewis Gordon makes this claim as part of his affirmation of black identity in what he calls "antiblack" racist society:

> What the Human Genome diversity Project reveals is that all human genes originated from the same region: Southeast Africa. During the period of our evolution, when human beings were in a single region, conditions were ripe for the maximum diversity of our gene pool. Subsequent patterns of movement and mating selections led to the focus on certain combinations of those genes in certain regions over others until the gene variations spread across the planet. What this means is that, from a genetic point of view, there is indeed one human species that originated from a single region. But here is the rub: race critics have read this conclusion to mean that races do not exist. In one sense it is true; races do not exist. *One* race exists, and that race is "Negro."[32]

Gordon is here speculating that non-black racial traits evolved from a totality of racial traits in an original black population. But, until it is independently established on empirical grounds that human races exist, i.e., that there are *racial* traits, there is no basis on which to project race onto an original African population, as Gordon does. The original African population may not have had the traits that later came to be associated with all races, or even the traits associated with the presumed black race. There is at present no way to confirm this one way or the other because, as noted earlier, the physical traits associated with race, such as skin color and hair texture, are formations of "soft tissue" for which there is no ancient archeological record.

There is no reason to believe, as Gordon does—and he has a huge cohort—that geographical distinction alone is sufficient evidence of racial distinction. The Human Genome Diversity Project and other endeavors in molecular genetics do not identify human geographical origins based on what are commonly taken to be physical racial traits, but based on genetic markers, which do no more than accompany population descent so that migrations, and the age of their resulting genealogical lineages, can be hypothesized. Thus, suppose that population B, on continent Q, is D genetic distance from population A on continent P. Suppose, also, that the ancestors of B were members of A, who traveled from P to Q and that no one in B interbred with members of A who remained on P after the Q-destination ancestors of B left. The number of years that it would take B to develop D genetic difference, or "distance" from A, is hypothesized to be the number of years since some members of A left P for Q and formed the new lineage. For out-of-Africa anthropologists, that kind of hypothesis is the only "result" of the available data, and it says nothing about race or racial taxonomy. If the multiregionalists are right, then, in the above model, there would have been gene flow between A and B after some As left P for Q, and that would leave less reason to posit distinctive racial traits in connection with distinctive geographical locations.

The origin of all modern *Homo sapiens* in Africa is consistent with the later development of distinct races, or with no racial development. It cannot be claimed, as Gordon does, that the common human origin is proof that there is one human race, namely the black one. Not only is geography alone insufficient to establish the existence of race, but "race" means a taxonomy or division into races. The existence of race is the existence of races, of groups generally distinct from each other, on the order of subspecies. If a species were divided into subspecies and all but one became extinct, it could then be said that the species presently had one race. But, this would be true only because of a prior existence of more than one race. Where there is one race from the beginning, there are no races. Gordon begs the question of whether there are human races. He extracts one component of commonsense racial taxonomy, namely "Negro," and simply projects it onto the scientific description of human origins. This does no more than reinscribe the common sense association of race with geography, which association is itself in need of a scientific foundation, onto a hypothesis about human origins that is not a hypothesis about racial taxonomy. (On top of this projection, Gordon seems to believe that an original "one drop" of African essence has persisted generationally for over 100,000 years![33])

Scientifically oriented mirages of race, based on geography, also now occur in the literature. Cavalli-Sforza's work on population genetics reconstructs the history of human geographical migration out of Africa in a way that can be modeled through geographical mapping and genealogical branching. Data on linguistic differences between populations can also be modeled in this way, given similar assumptions about migrations and descent, as are made with populations biologically. And, the linguistic mapping and branching appears to coincide with the linkage based on biology.[34] The substantial coincidence of successive geographical populations with commonsense continent-based racial categories may seem to fulfill a long promise of a a scientific foundation for those categories. But how, exactly?

Robin Andreasen considers the taxonomies of migratory populations that for long periods of time did not interbreed with other populations or with the descendants of their ancestral populations. She suggests that branches of such populations be viewed as *clades*. A clade is an ancestor and descendants who form an isolated breeding population, which in evolutionary biology can be used as the unit in models of taxa higher than species. Thus, Andreasen uses the concept of clades to interpret Cavalli-Sforza's data as support for a scientific concept of race.[35] The problem with this proposal is that there is no empirical reason to use Cavalli-Sforza's data as a basis of the commonsense concept of race, because his data is based on genetic differences that do not have any known biological relation to those traits that in common sense are associated with race. Cavalli-Sforza and his colleagues selected original subjects for DNA mutational comparison based on a pre-genetic, or social, classification into black and white racial groups in England and the United States, and membership in over forty geographically diverse aboriginal groups. The black and white individuals were assumed to have ancestral origins in Africa and Europe, and the aboriginal groups were assumed to have been in their present abodes for long periods of time.[36] Thus, well-founded presumptions about geographical origins were built into the selected categories whose geographical ancestry were to be tracked. But where does race come into this scientific picture, if the commonsense racial taxonomy is merely assumed and the geographical origins are not in themselves racial? For instance, if all the people identified as white had ancestors alive in Europe during the same time that the people identified as black had ancestors alive in Africa, to say that these are racial ancestral differences adds no new information to the data on time and place.

Cavallli-Sforza's preselection was made in order to establish genetic base points from which mutations could be identified. The genetic mutational data does not identify race scientifically, but merely identifies likely geographical origins of groups (i.e., of the ancestors of selected groups) and paths of population migration. Andreasen is aware of Cavalli-Sforza's own resistence to racialist (i.e., entailing belief in the existence of races) interpretations of his data, but she fails to offer an argument against his resistance. The only way Andreasen's races-as-clades proposal could provide a scientific foundation for commonsense racial taxonomy is if it is assumed that geography in itself grounds race, so that different ancestral places, even if occupied for only 40- or 50,000 years, automatically entails the existence of different races. Skepticism about a scientific foundation for race does not include denial of the ability to prescientifically identify those groups that are presumed to be races—what's at issue is whether the groups are races in any scientific sense. Furthermore, confirmation of the multiregionalist hypothesis would automatically exclude a notion of races as clades, because there would be little evidence for the reproductive isolation of human groups. Indeed, if the multiregionalists are correct, no human group that was smaller than the entire species could be a clade.

The question remains open of why geography should be assumed to cause race or how geographical differences could cause racial differences. A scientific basis for the common assumption that geography causes races would require either a causal explanation of how general or specific racially salient physical traits were linked to geographical conditions, or, coherent statistical support for the conjunction of general or specific, racially salient physical traits, with some geographical conditions and not others. Geographical origins of ancestors, alone, could not provide the causal or statistical account because it is the hereditary effects of geography over time that are supposed to do the racial work. That is, geography is commonly assumed to account for racial differences "as a result of evolution." But this is vague. If Homo sapiens has evolved over the past 100,000, 200,000, or two million years, adaptive changes are the evidence for it. But what is the evidence that significant and uniform adaptive human changes are correlated with differences in races?

The main apparent difference associated with race, skin color, was until recently believed to be the result of natural selection within populations living under different amounts of sunlight. However, dark-skinned groups live in cold, as well as warm, climates. The hypothesis that dark skin increases "fitness" in sunny climates, because it offers protection against ultraviolet rays that can cause skin cancer, is undermined by the fact that skin cancer rarely develops in individuals before they are of reproductive age. (That is, in terms of natural selection, fitness means increased reproduction due to an adaptive trait and not the adaptiveness of individuals independently of whether of not they reproduce.) The hypothesis that light skin facilitates the synthesis of vitamin D in climates with little sunlight is weakened by the fact that there is no evidence of vitamin D deficiency (rickets) in past European populations when winters were more severe, and also by recent evidence that vitamin D can be stored in the human body for the length of a northern winter.[37]

However, before skin color differences can be used as evidence for a geographical basis of race, it needs to be independently shown that skin color differences themselves form a taxonomy that corresponds to common sense notions of race. The account of why that cannot be done exceeds the scope of this paper.[38] Here, it can be concluded that the geographical account of human history in itself yields no evidence that modern Homo sapiens has evolved in different ways based on groupings that can be defined in racial terms. To repeat, geographical origins do not in themselves constitute races and to assume that they do, in the absence of comprehensive supporting human evolutionary data, is an egregious bit of flim-flam, which begs the question of whether there are races. Overall, the geographical basis of human racial taxonomy seems to be no more than a rhetorical tradition deriving from Eurocentric reactions to contact with inhabitants of other places. Descriptions of how the inhabitants of Asia, Africa, and America appeared to Europeans during early modern contact are fascinating (and painful) to read, and genetic models of human population migration out of Africa are interesting in their own right as accounts of our species history. But neither type of account yields any more scientific information about race than Amerigo Vespucci's 1503 *Mundus Novis*.

Acknowledgment

This essay first appeared as Chapter 2 of Naomi Zack, *Philosophy of Science and Race* (Routledge, 2002). We thank Routledge for permission to reprint.

Notes

1. Thanks to John Relethford for making the explicit connections expressed in the last two sentences of this paragraph in his review comments.
2. Ashley Montagu, *The Idea of Race* (Lincoln, Neb.: University of Nebraska Press 1965), 81–116; idem, *The Concept of Race* (London: The Free Press of Glencoe, Macmillan, 1994), xii–xviii.
3. For a useful collection of eighteenth and nineteenth century theories of white supremacy, including writing by Thomas Jefferson, Comte de Buffon, Hume, Kant, and Hegel, see Emmanuel Chukwudi Eze, ed. *Race and the Enlightenment: A Reader* (Cambridge, Mass.: Blackwell Publishers), 1997.

4. See Stephen Jay Gould, *The Mismeasure of Man* (New York: W.W. Norton, 1996).

5. Claude Levi-Strauss, "Race and History," in *Race, Science and Society*, Leo Kuper, ed. (New York: Columbia University Press, 1965), 95–134.

6. Amerigo Vespucci, *Mundus Novis*, reprinted in *Philosophy in the 16th and 17th Centuries*, Richard H. Popkin, ed. (New York: The Free Press, 1966), 26.

7. George-Louis Leclerc, Comte de Buffon, *A Natural History, General and Particular*, reprinted in Eze, *Race and the Enlightenment*, 20–21.

8. Johann Friedrich Blumenbach, "Degeneration of the Species," from *On the Natural Varieties of Mankind*, reprinted in Eze, *Race and the Enlightenment*, 82–3.

9. Georg Wilhelm Fredrich Hegel, "Geographical Bases of World History," from *Lectures on the Philosophy of World History*, reprinted in Eze, *Race and the Enlightenment*, 148–9.

10. Ibid., 124.

11. Daniel G. Blackburn, "Why Race Is Not a Biological Concept," in *Race and Racism in Theory and Practice*, Berel Lang, ed. (Lanham Md.: Rowman and Littlefield, 2000), 3–26, especially 13–14.

12. Richard R. Popkin, "Hume's Racism," in *Philosophy and the Civilizing Arts: Essays Presented to Herbert W. Schneider*, Craig Walton and John P. Anton eds. (Athens, Ohio: Ohio University Press, 1974), 150–1.

13. See Milford Wolpoff and Rachel Caspari, *Race and Human Evolution* (New York: Simon and Schuster, 1997), 136–172.

14. See ibid., and Milford H. Wolpoff, John Hawks, and Rachel Caspari, "Multiregional, Not Multiple Origins," *American Journal of Physical Anthropology* 112 (2000): 129–136. See also Blackburn, op. cit.; John Noble Wilfred, "Skulls in Caucasus Linked to Early Humans in Africa," *New York Times*, May 12, 2000, A1.

15. Luigi Luca Cavalli-Sforza, *Genes, Peoples and Languages* (New York: Northpoint Press, 2000), 57–66.

16. Blackburn, op. cit., 18.

17. See Relethford, "Models, Predictions and the Fossil Record of Modern Human Origins," *Evolutionary Anthropology* 8 (1999):7–10.; Wolpoff and Caspari, *Race and Human Evolution*, especially 257–9.

18. Cavalli-Sforza, *Genes, Peoples and Languages*, 61–6; See also Richard Preston, "The Genome Warrior: Craig Venter's Race to Break the Genetic Code," *The New Yorker*, June 12, 2000, 66–77.

19. John H. Relethford, *Genetics and the Search for Modern Human Origins*, (New York: Wiley-Liss, 2001), 194–211; Lynn B. Jorde, Michael Bamshad, and Alan R. Rogers, "Using Mitochondrial and Nuclear DNA Markers to Reconstruct Human Evolution," *BioEssays* 20 (1998):126–136.

20. Relethford, however, who is a multiregionalist, thinks that skull proportions, although not skin reflectance may be a reliable index of population history, because craniometric variation among modern human populations, while showing a low variation among groups, varies about the same amount among groups as do genetic markers. (See Relethford, "Craniometric Variation Among Modern Human Populations," *American Journal of Physical Anthropology* 95

(1994): 53–62.) However, it is difficult to understand how a statistical correlation between genetic marker differences and craniometric indices could mean that craniometrics is a reliable index of population history, without either an independent phenotypic basis for population taxonomy that privileged craniometrics, or an explanation of some causal link between genetic markers, which do not influence somatic development, and craniometrics. Furthermore, as noted, on the multiregional hypothesis, there is little reason to think that any population has a single ancestry, on account of gene flow.

21. For a discussion of the problems with a phenotypical basis of race, see Naomi Zack, *Philosophy of Science and Race* (New York: Routledge, 2002), ch. 3.
22. Cavalli-Sforza, op. cit.
23. Ibid.
24. Ibid.
25. Jorde et al., "Using Mitochondrial and Nuclear DNA Markers to Reconstruct Human Evolution," *BioEssays*.
26. Cavalli-Sforza, *Genes, Peoples and Languages*, 77–82.
27. Nicholas Wade, "The Human Family Tree: 10 Adams and 18 Eves," *New York Times*, May 2, 2000, F1.
28. Cavalli-Sforza, *Genes, Peoples and Languages*, 82–5.
29. For different versions of multiregional and replacement, or out-of-Africa theories, see: Relethford, *Genetics and the Search for Modern Human Origins*, 67–81; Jorde et al., "Using Mitochondrial and Nuclear DNA Markers"; Alan R. Rogers and Lynn B. Jorde, "Genetic Evidence on Modern Human Origins," *Human Biology* 67 (1995):1–36.
30. See note 20, above.
31. An interesting anachronistic theme in this regard is expressed in Charles Mills's rhetorical attempt to interpret modern European hegemony as a "racial contract," by projecting later biological notions of race onto early colonialist and merchantilist ideologies and political philosophies. Mills attempts to use European and American social contract theory (from Locke to Rawls) to explain Western white supremacy. However, the victims and objects of colonialism did not consent to their enslavement and exploitation in the informed way that social contract theory would require, and there is no trace of irony, or even sarcasm, in Mills's polemics on this score. (Charles W. Mills, *The Racial Contract*, Ithaca, N.Y.: Cornell University Press, 1997.)
32. Lewis Gordon, *Existentia Africana* (New York: Routledge, 2000), 82–3.
33. Gordon goes even further than this with his claim that African Americans have "long been able to see" that the physical traits of some whites indicate their African ancestral origins and thereby vouchsafe their "Negro" identities. Gordon believes that all human genetic diversity existed in the original African population and that through selective breeding, that population developed into apparently different races. He thinks that, nonetheless, the group designated "Negro" today has greater genetic diversity than any other group. One wonders how Gordon can say both that everyone is black and that the group designated as black is different from other groups in diversity, but that contradiction could be overlooked had Gordon not overlooked, or failed to notice, the presently well-publicized fact that any small human population, anywhere

on earth, now contains most of the diversity in the human genome (see Chapter 4 in Zack, *Philosophy of Science and Race*, for a discussion of the last; see ibid., p. 83 for what I have here attributed to Gordon). There is also a moral dimension to Gordon's version of population genetics. He says that blackness is synonymous with humanity. (Ibid.) Gordon seems not to have advanced scholarly discussion of race and science beyond the place secured when Eldridge Cleaver retold Elijah Mohammad's myth about Dr. Jacob, a black scientist who bred "the white devil with the blue eyes of death" from an original, (angelic?) black population. (See Eldridge Cleaver, *Soul on Ice*, New York: Dell, 1968, 99) Such rhetoric, no matter how self-intoxicating or otherwise therapeutic it may be, does no more than switch the poles of insult from white to black and black to white, without, sadly, testing the ground supposed to support them.

34. Cavalli-Sforza, *Genes, Peoples and Languages*, ch. 5. See also, idem. and Francesco.

35. Robin O. Andreasen, "A New Perspective on the Race Debate," *British Journal of Philosophy of Science* (1998) 49: 199–225.

36. Cavalli-Sforza, *Genes, Peoples and Languages*, 57–85.

37. See Alain Corcos, *The Myth of Human Races* (East Lansing: Michigan State University Press, 1997), 83–8.

38. See Zack, *Philosophy of Science and Race*, ch. 3.

Queer Nature, Circular Science

Margaret Cuonzo

*Phenomena such as homosexuality or gender mixing are never
seen as neutral or expected variations along a sexual and gender
continuum (or continua), but rather as abnormal or exceptional
conditions that require explanation.*

—Bruce Bagemihl, *Biological Exuberance*

Introduction

Queer theorists, scientists, feminists, philosophers of science, and racial
theorists alike have used the time-honored critical method of showing that
the evidence given in support of a conclusion assumes the conclusion itself.
Feminists, for example, criticize certain sociobiological positions on gender relations for appealing to evidence that is amassed in a way that assumes those very same gender relations.[1] Queer theorists can, and have
used this approach to arguments concerning the prevalence and natural-
ness of homosexuality and other forms of nonheterosexuality.[2] To the best
of my knowledge, though, no queer theorist or any type of theorist mentioned above has given a detailed analysis of the ways of making this kind
of critique. This is one of the two main goals of this article. First, I present a
new way of looking at circularity in scientific studies. And second, I apply
this analysis of the two types of circularity to studies concerning animal
sexuality. There are at least two ways of using circular reasoning in science,
what I have termed *descriptive* and *normative* circularity and analyze below.
And scientific studies about animal sexual behavior provide interesting examples of these two forms of circularity and of scientific bias against nonheterosexuality. To accomplish the second goal, I have referred primarily,
but not exclusively, to the most thorough account of animal nonheterosexuality and critique of scientific methods of amassing data regarding animal

sexuality, Bruce Bagemihl's massive (1999) *Biological Exuberance: Animal Homosexuality and Natural Diversity*. Although Bagemihl is by no means the first to raise issues regarding members of the scientific community's treatment of animal sexuality, his is a recent and thorough account that is worthy of discussion. In addition, it provides ample evidence for constructing an account of two ways that circularity leads to bad science. And although Bagemihl is not primarily concerned with constructing a more general methodology for doing science, and I have reservations about the moves he makes in this direction, we can use his informative and interesting book to help construct such a methodology.

Overview of Two Types of Circularity

As mentioned above, the criticism that can be made of certain types of reasoning about sexuality, what I call the *circularity critique*, takes issue with two forms of circularity. The first form, *descriptive circularity*, is the following: certain scientific studies amass data about nonhuman animals in a way that assumes that there are no, or few, natural occurrences of nonheterosexual sexual behavior in these animals. As a result of this, they offer as scientific "truth" studies that radically misrepresent the natural world and license the claim that there are no, or few, occurrences of nonheterosexuality in the natural world. Finally, in another fateful inference, these studies (either directly or indirectly) license the conclusion that nonheterosexuality in humans is therefore unnatural, due to its absence in the rest of the animal kingdom.

The second form of circularity is called *normative circularity*, although both forms have both normative and descriptive elements. To be circular in this second way, the scientist begins with the assumption that nonheterosexuality is immoral, perverse, deviant, and so on. That is, the scientist takes some kind of normative stance on nonheterosexuality in general[3]. Then, when confronted with instances of nonheterosexual behavior in nonhuman animals, he or she concludes that the animals are exhibiting some kind of deviance, or perversion. And again lastly, the "perversion," or "deviance," of creatures in the animal kingdom, is (indirectly or directly) offered as support of the deviance of nonheterosexuality in humans. This last inference is not as common, because the obvious circularity is hard to go unnoticed. So what usually is employed is a combination of the two forms of circularity, that is, the claim that *the few* instances of nonheterosexuality in the animal world are instances of deviance or perversion, and therefore, the instances of human nonheterosexuality are to be classified in the same way.

Descriptive Circularity in Action

Descriptive circularity occurs when evidence is amassed in a way that already assumes a conclusion that will be eventually drawn from the data. With regards to certain biological studies of animal sexuality, the data amassed about animal sexual behavior were gathered in a way that ignored, or that otherwise discounted, nonheterosexual behavior. An example of such a method occurs in studies that use behavior-based methods for the determination of sex. Behavior-based methods involve defining the sex of otherwise indistinguishable animals in terms of their behavior during sex. For example, in a 1975 study of laughing gulls,[4] the following method was used to determine the sex of the gulls: (i) if a bird was on top during copulation more than two times, then it was determined to be a male; and (ii) the mate of the *male* was presumed to be female. On this method, there is no possibility for concluding that there are any nonheterosexual instances of copulation. This method also precludes there being an instance of sexual interaction where a female is the bird on top of a male. Unfortunately, even in species where there are a significant number of well-documented cases of nonheterosexual behavior, such methods continue to be used. The problems with the assumptions made by studies such as these are twofold: (a) the study misrepresents reality in a way that is potentially relevant to the conclusion drawn by the scientist; and (b) assumptions such as these lead to the collective form of descriptive circularity, whereby a conclusion is drawn, perhaps by someone else, that there are no instances of nonheterosexual behavior in laughing gulls.

Another way to amass data in a way that assumes a conclusion that will be eventually drawn is simply to avoid reporting relevant data. In our case, this involves not reporting cases on nonheterosexuality. This then sets the stage for concluding that since such behavior in the rest of the animal world "does not exist," it is abnormal for such behavior to exist in humans. For example, in a 1991 article, primatologist Linda Wolfe writes the following about her discussions with many primatologists and their reluctance to report data:

> I have talked with several (anonymous at their request) primatologists who have told me that they have observed both male and female homosexual behavior during field studies. They seemed reluctant to publish their data, however, either because they feared homophobic reactions ("my colleagues might think that I am gay") or because they lacked a framework for analysis ("I don't know what it means") (30).

Wolfe's observation describes two ways in which data can be amassed in a suspect way. The first, more questionable, way is to acknowledge the presence and importance of such data, but not report the data. The second way is not to publish the data because the scientist does not know how the data are to be interpreted. For example, in the following passage, a scientist studying rams describes his own eventual awareness that the mounting behavior between male bighorn rams was sexual:

> I still cringe at the memory of seeing old D-ram mount S-ram repeatedly. . . . True to form, and incapable of absorbing this realization at once, I called these actions of the rams *aggressosexual* behavior, for to state that the males had evolved a homosexual society was emotionally beyond me. To conceive of those magnificent beasts as "queers"—Oh God! I argued for two years that, in sheep, aggressive and sexual behavior could not be separated. . . . I never published that drivel and am glad of it. . . . Eventually I called a spade a spade and admitted that rams lived in essentially a homosexual society.[5]

Perhaps because of the negative stereotypes associated with nonheterosexuality in humans, this scientist was at first not able to admit that the behavior of the rams was sexual. By not admitting to such cases, the studies that are eventually published misrepresent the subjects being studied. As a partial consequence of such misrepresentation of data, standard biological textbooks often do not mention occurrences of nonheterosexuality in animals. In fact, even in cases where there are published findings about nonheterosexuality in animals, the information is often left out of introductory texts. In 1995, for example, primatologist Paul Vasey wrote, "Although the first reports of homosexual behavior among primates were published 75 years ago, virtually every major introductory text in primatology fails to even mentioned its existence."[6]

Another way to avoid publishing data is to set a significance level for the occurrence of a certain behavior high enough so that what is usually regarded as a significant number of occurrences is regarded as insignificant. In science, a standard, though arbitrary, way for defining significance is to say that if something happens more than five percent of the time, then it is significant. Yet, when it comes to nonheterosexuality in scientific studies, the significance of occurrences that happen at far higher percentages is called into question. And even when it is recorded, the data are downplayed. Consider for example, the following investigation of lesbian pairs of Western gulls: "We have estimated female-female pairs make up only 10–15 percent of the population."[7] If the standard level for significance is 5 percent, then 10 to 15 percent of a given population is not something to preface with the term "only." Even if the author of this study is merely giving an informal estimate, a quite significant activity that would potentially shed light on the sexual activity of the gulls has been put to the side. Thus, even interpreting the author of the study in the best light possible, there is a mistake in discounting 15 percent of a behavior as irrelevant.

Many of the studies that are descriptively circular do not have as their main conclusion that there are no instances of nonheterosexuality in the animals they are studying. The main conclusions may be about the mating behavior of the animals, parenting, and so forth. Yet, the main conclusions of the findings imply that there are no, or few, instances of nonheterosexuality in the animals being studied. In some cases, the data are presented in a way that logically entails that there are no, or few, instances of nonheterosexual behavior in the animals being studied. Consider the behavior-based method of sex determination. It is not logically possible to find nonheterosexual behavior using this method. And, although the method of not reporting data does not logically entail that such instances do not occur out of view of the scientists, the lack of such data provides strong inductive support that such behavior does not exist. If the vast majority of studies about animal sexuality do not report a certain occurrence, then it is quite reasonable to conclude that such behavior does not exist. While the studies discussed earlier do not have overtly political or polemical conclusions, the data are amassed in a way that gives "evidence" for others with political agendas to draw on. By looking through the literature regarding animal sexuality, one might easily conclude that nonheterosexuality in nonhuman animals is quite rare, and that therefore, human nonheterosexuality is somehow "unnatural," or as talkshow host Dr. Laura Schlessinger claimed, "a biological error."[8] This brings to light a strange property of this type of circularity; it is often, though not exclusively, committed collectively. A study has elements that are overtly or covertly biased against nonheterosexuality, and then someone else, possibly in a far different place and time, uses these "findings" to draw conclusions about the "unnaturalness" or "abnormality" of non-heterosexuality, as in the Dr. Laura case.

Let me summarize the descriptive circularity involved in studies involving behavior-based sex determination and the selective reporting of data. By collecting data in these ways, certain studies already assume that there will be no, or few, instances of nonheterosexuality in the animals they study. The data then published as scientific "fact" license the conclusion that there are no, or few, such instances. In addition, those with a political/religious stance on nonheterosexuality with regard to humans can use this "scientific fact" to further their own political/religious agendas.

Normative Circularity in Action

I have distinguished descriptive circularity from normative circularity, because descriptive circularity does not take an overtly normative stance on animal behavior. It simply creates a system of reporting that excludes non-heterosexual behavior. Surely, there is some normative stance taken by such studies regarding nonheterosexual behaviors, but the studies, in not recording the data, do not label the behavior of the animals in normative terms. The descriptive kind of circularity simply avoids saying anything about nonheterosexuality, at least explicitly. The second type of circularity, normative circularity, begins with a scientist taking a normative stance on some behavior generally, such as in our case, nonheterosexuality. The scientist then describes the behavior of the animals that are behaving in ways that are nonheterosexual in these terms. In other words, the animals' behavior is described as "perverse" or "abnormal," along with other negative value-laden terms.

One of the most blatant examples of this type of reasoning is a surprisingly recent (1987) article titled "A Note on the Apparent Lowering of the Moral Standards in the Lepidoptera."[9] The lepidoptera are Mazarine Blue butterflies, and the article concerns the homosexual mating of these creatures. The author waxes "philosophical" at the beginning of the discussion: "It is a sad sign of our times that the national newspapers are all too often packed with the lurid details of declining moral standards and of horrific sexual offences committed by our fellow Homo sapiens; perhaps it is also a sign of the times that the entomological literature appears of late be heading a similar direction." After this, the author discusses his findings concerning the nonheterosexuality observed in these butterflies. This is a clear case of normative circularity, because the author has already indicated, by virtue of the prefatory comments, a normative stance on the issue of non-heterosexuality. Since nonheterosexuality, in principle for Tennent, the author, is a lowering of moral standards, the butterflies that engage in such behavior are being immoral. Of course there is a problem in discussing the behavior of butterflies in moral terms, since this assumes that the butterflies have intentions, conscience, or awareness of morality, but I'll put this aside in order to focus on the circularity of the reasoning. Some might think that the author of the article was simply making a joke, with no assignment of moral values to the butterflies. However, the prevalence of this type of discussion in the scientific literature indicates that there is something far more serious at work. Here are just a few of the titles that could be interpreted in a similar way: "Aberrant Sexual Behavior in the South African Ostrich," "Abnormal Sexual Behavior of Confined Female *Hemichienus auritus syriacus*," and countless others.[10] Here is another surprisingly recent description of this value-ladenness in a 1982 description of waterbuck: "Among aberrant sexual behaviors, anoestrus does were very occasionally seen to mount one another" (Spinage 1982, 118). Clearly, the author is taking a normative stance on such behavior.

As with the studies that commit descriptive circularity, the next step is to present the scientific "findings" to the general scientific community. The "deviance" of the animals in the studies then licenses the conclusion that nonheterosexuality in humans is likewise deviant. Usually, though, the claim is offered in conjunction with the descriptive claim that there are few such occurrences. Consider the report about the does, which were "very occasionally seen to mount one another." The conclusion derived from the normative circularity is that the few occurrences of nonheterosexuality are cases of "deviance" or "abnormality." Like the descriptive form of circularity, the relevant form of normative circularity is often committed collectively. The first step is to apply a normative stance on nonheterosexuality to the animal world. The next step, often made by another person, is to take the "findings" from the first step and then conclude that human behavior, like its correlate in the animal world, is worthy of being characterized in the same way.

The criticism that I am making of the normative form of circularity with respect to animal sexuality is strikingly similar to two more famous critiques of scientific theories, taken together: Marx's critique of Charles Darwin's theory of evolution, and the criticism of social Darwinism. Although a Darwinist in certain respects, Marx, in his letters, criticizes Darwin as misrepresenting the animal world as behaving in a way that was similar to the behavior of humans in the Victorian England in which Darwin lived. And social Darwinists are criticized for then using his "findings" regarding the animal world to give an account of the behavior of our own species. Similarly, I am alleging that scientists such as Tennent and Spinnage commit the error that Marx claims Darwin made.[11] The error is that of inferring falsely from claims about human interactions to claims about butterflies and waterbuck. Tennent and Spinnage have interpreted their subject's behavior in moral terms because they interpret similar behavior in humans in those terms. This interpretation of the behavior of butterflies and does, combined with the supposedly rare occurrence of such behavior, again licenses the conclusion that human nonheterosexuality is a "lowering of moral standards," or "aberrant."

Methodological Considerations

Unfortunately, I do not have a surefire method of avoiding descriptive and normative circularity. However, the definitions of the two kinds of circularity do suggest some ways to avoid making these kinds of mistakes. Descriptive circularity occurs when a conclusion is drawn by amassing data about the existence of certain behaviors in a way that would not allow there to be another conclusion drawn. In the case of behavior-based sex determination, for example, it is not possible for there to be an instance of non-heterosexuality. Such a method assumes that only heterosexual sex acts exist or matter, and hence the "findings" will discuss only those behaviors. Thus, one way to avoid descriptive form of circularity is to treat all possible behaviors within certain well-defined limits as relevant to the study. The problem, though, is to find well-defined limits for the study.

Perhaps an historical example of how a method of data collection was criticized for being biased in this way would be helpful. In 1974, Jeanne Altmann published a paper titled, "Observational Study of Behavior: Sampling Methods"[12] that criticized methods for studying primates and proposed another method for the collection of data. At the time in which Altmann wrote, studies of primates focused heavily on violence and sex to the exclusion of other activities. Altmann criticized the methods used for sampling data, such as the "one-zero" sampling method, where the occurrence and nonoccurrence of an event was recorded. A scientist could use the one-zero method to analyze how often an event, for example an act of aggression on the part of an animal, occurred. This was done by observing the animal and then recording an act of aggression as a "1" and the absence of such an act as a "0." Such a method, according to Altmann, has no use in a scientific study. In the case of descriptive circularity discussed in this article, it is easy to see potential problems with this method. If a scientist is questioning whether an animal is engaged in sexual activity, she or he might be inclined to assign a "0" to behavior that is sexual on the basis that the animals involved are of the same sex. Moreover, the very types of questions formulated could skew an objective picture of the animals being studied. If all the questions asked about the animals concern sex and violence, then the wide body of literature on them will paint a very narrow picture of the animals.

As a replacement method, Altman proposed what she termed the "focal point sampling" of data. According to Donna Haraway, such a method "could result in data appropriate for examining sequential restraints on particular kinds of events or conditions, percent of time spent doing something or being in a given state, as well as rates, durations, and nearest neighbor relations."[13] Using this method, all events are recorded for a certain pre-set time. Such a method could be used to answer comparative questions about groups being studied and is an improvement over the one-zero method of data collection. The focal-point sampling method of data collection is an example of a way in which bias in science can be minimized. Such a method would note the specific activities of animals within a specific time period rather than looking for certain activities that were already anticipated by the scientist. In this sense, it is more objective. Such a method, for example, would record data regarding the interaction of animals such as stroking, and so on, rather than anticipate behaviors according to a preconceived idea about what counts as instances of sexual activity. However, focal-point sampling is not, nor was it intended to be, a surefire method for avoiding bias and human error in science. Although the potential for descriptive circularity can be lessened, it is not eliminated.

Normative circularity is not much easier to avoid. The strategy here is to avoid taking a normative stance on behavior. Terms like "ought," "should," "abnormal," and "immoral," should be severely limited when drawing conclusions about nonhuman animals.[14] Since mechanical methods of data collection record events without assigning value judgments, such methods may be useful. For example, an infrared camera could simply record behavior without labeling it as "immoral" or "abnormal." However, the problem here is that the data must then be interpreted, and this is where the question of value-laden observation arises. Since none of these methods completely avoid circularity, we are left with an account of two types of circularity and a few suggestions on how to avoid them.

Here one might respond that normative and descriptive circularity are not the only ways in which conclusions about nonheterosexuality are drawn, and that the most blatant form of bias against nonheterosexuality comes from religious circles. Why worry about these two ways of doing bad science? I have two answers to this question. First, the purpose of "worrying about the scientists" is that we can see two forms of bad inference in action, and perhaps can learn from these errors. The second is that, regardless of how influential or not such studies are, they can be, and have been, used in suspect ways, and this needs to be pointed out.

Such a method cannot, and was not intended to, completely avoid bias and human error, but it can minimize the potential for bias.

Some Deeper Questions

Although my primary aim is to critique the reasoning involved in various accounts of animal sexuality, doing so only proves that bad science was taking place. Such a critique does not show that the conclusions drawn are false. Positive evidence is needed to prove the falsity of the conclusion. So, now I'll briefly outline evidence about the nature of animal nonheterosexuality. Readers who would like a further information can refer to Bagimihl (1999) themselves. According to Bagemihl, same-sex sexual acts, in addition to same-sex courtship behavior, pair-bonding, and parenting have been observed in 471 species. This includes 167 species of mammals, 132 birds, 32 reptiles and amphibians, 14 fishes, and 125 insects and other invertebrates.[15] The frequency of these behaviors varies depending on the species, from less than five percent of the overall sexual activity in herring gulls to over 50 percent in male giraffes.

To give some examples, there is a fair amount of evidence concerning nonheterosexual behavior in farm animals. Animals that have been reported to engage in such behavior include: cattle, sheep, goats, pigs, and horses of both sexes that mount same-sex partners. Same-sex pair bonds have been observed in pigs, sheep and goats. Indeed, due to the prevalence of such activity, there are special terms coined by farmers and breeders to refer to the same-sex sexual behavior of the animals. As Bagemihl writes, "mounting among male Cattle is referred to as the "buller syndrome" (steers who are mounted are called "bullers," the males who mount them are called "riders"), female sows who are mounting each other are described as "going boaring," mares who do so are said to "horse," while cows are said to "bull." In addition, in breeding programs, homosexual relations are often used to promote heterosexual relations among the animals. In some species, for example, female-female mounting is a reliable indicator of when the females are in heat. Also, young bulls or steers, which are called "teasers," are often presented to the mature bulls in order to arouse the bulls and facilitate semen collection for artificial insemination.

Here though, some deeper questions emerge about a correct method for characterizing the behavior of nonhuman animals. For example, in Bagemihl's book there are many pictures of animals in what looks like sexual activity. In each of the pictures, there is a pair of animals of the same sex performing what seems to be a sex act. But how do we know that these behaviors are what they seem to be? We do not often ask these questions of different sex animals, but does that mean the question is any less telling? That is, we *could* and perhaps *should*, ask this question concerning different-sex animal pairs as well. In fact, we can even ask this concerning human animals engaged in what seems to be a sex act. How do we know what is going on? The behavior pictured could be aggressosexual behavior, as the scientist I quoted previously believed. Such questions recommend a far more tentative approach to the behavior of all animals.

The troubles associated with providing an accurate interpretation of animal, and even human, sexual behavior can be seen as an instance of a more general problem, the "other minds" problem. It is famously difficult to conclusively ascribe specific intentions, beliefs, and so forth, to other human beings, let alone to animals. It is even difficult to show conclusively that other creatures have mental states at all. Yet, it is a precondition on knowing that some interaction is sexual (and not agressosexual, grooming, etc.) that we know that the act involves certain states within the animals' minds and bodies, such as specific sensations or desires. Therefore, labeling any type of behavior sexual must, to varying degrees, be problematic. Because it is preconditioned on being a sexual act that the actors be in certain mental states and have certain intentions, and because these mental states are at least to some degree unavailable, there will always be a potential lack of fit between our conclusions about the sexual behavior in others and the actual behavior.

In addition to the epistemological problems associated with knowing the intentions of other creatures, there are additional problems associated with the meaning of the phrase "sexual behavior." In the previous paragraph, I assumed that one necessary condition in an act being sexual is that the actors have certain inclinations towards the act. Yet, to be more specific leads to difficulty. There are a range of states, intentions, and actions that can be lumped into the broad category of "sexual behavior," making it difficult to give a precise definition of "sexual behavior." For example, if "sexual behavior" is defined as "actions using the sexual organs," the question of circularity again arises. Why are these organs sexual? If "sexual behavior" is defined as "actions leading to reproduction," the definition leaves out nonreproductive sexual behavior. If "sexual behavior" is defined as "actions with an intent to lead to orgasm," there are two problems: first, the problems associated with saying that nonhuman animals have intentions, and second, the problem with sexual actions that are not intended to necessarily lead to orgasm. What this suggests is that the ordinary language notion of sexual behavior is problematic.

Just as it is possible and, in fact, the case that there is a heterosexist bias in scientific accounts of animal sexuality, there is also the potential for homosexist, or more likely, anthropocentric bias, or even egocentric bias. If, as the opening quotation to Chapter 1 of Bagemihl's book claims, the world is not only queerer than we suppose, but queerer than we can suppose, our conclusions must remain tentative.[16]

Notes

1. See Fausto-Sterling, 1995. See also Lloyd, 1996.
2. I will use the term "nonheterosexuality" as an umbrella term to refer to any instance of sexuality that does not occur between one male and one female crea-

ture. The term, as I use it, also applies to bisexuality, transgender relations, and so on. It is important to note that I am not using the term to refer to an inherent disposition to behave in a certain way. I am referring to events only.

3. By the way, it is possible to assume a positive moral stance and commit the same kind of error. To my knowledge, there are no such studies.

4. Burger and Beer. 1975: 35. Cited in Bagemihl, 1999. 94.

5. Geist. 1975. 97–98.

6. Quoted in Bagehmihl, 1999: 103.

7. Hunt et al. 1980. Cited in Bagehmihl. 1999: 101

8. Dr. Laura Schlessinger. 1998. ⟨http://www.DrLaura.com.⟩. Accessed December 8, 1998. It is important to note that the reference to Schlessinger was not used as an analysis of her work as a "scientist" but as an example of how those outside the scientific community can seize upon studies that exclude nonheterosexuality and, through collective descriptive circularity, conclude something about an activity that was assumed by the study of the activity. It is also important to note that, clearly, Schlessinger's use of the term "error" is not a good one. My point is not to highlight some feature of Schlessinger's comments but to illustrate a potential case of collective circularity.

9. Tennent, W.J. 1987: 88–89. See also Crews and Young 1991: 512–514.

10. See in Bagemihl 1999, 685.

11. I am not taking issue with the theory of evolution, but merely pointing out that Marx used a similar line of reasoning to criticize a scientific theory.

12. See also Haraway's chapter on Altmann in Haraway 1989.

13. Haraway, 1989, 306.

14. It is important to note that there is a distinction between "should" being used in a normative sense and in a probabilistic one. Used in the normative way, "should" deals with questions of morality, goodness, vice, virtue, and so on, such as in "You should not have killed him." Used in the probabilistic sense, "should" refers to the probability that something is the case, such as in "The planets should exhibit elliptical orbits." The "should" I am discussing is the former, normative variety.

15. Bagemihl, 1999, 673, note 29.

16. I would like to thank Sandra Harding, Robert Figueroa, the National Science Foundation, the LIU Faculty Released Time Committee, Elisabeth Lloyd, Janet Haynes, the LIU Biology Honors Society, and two anonymous referees for their suggestions and support.

References

Altmann, Jeanne. "Observational Study of Behavior: Sampling Methods." *Behaviour* 49 (1974): 227–67.

Bagemihl, Bruce. *Biological Exuberance: Animal Homosexuality and Natural Diversity.* New York: St. Martin's Press, 1999.

Burger, J. and C. G. Beer. "Territoriality in the Laughing Gull (1. atricilla)" *Behavior* (1975) 55: 301–320.

Crews, David and Larry J. Young. "Pseudocopulation in Nature in a Unisexual Whiptail Lizard," *Animal Behavior* 42 (1991):512–4.

Fausto-Sterling, Anne. *Myths of Gender: Biological Theories about Women and Men.* New York: Basic Books, 1985.

Geist, Valerius. *Mountain Sheep and Man in the Northern Wilds.* Ithaca: Cornell University Press, 1975.

Haraway, Donna. *Primate Visions: Gender, Race, and Nature in the World of Modern Science.* New York: Routledge, 1989.

Hunt, G.L.J., et al. "Sex Ratio of Western Gulls on Santa Barbara Island, California" *AUK* 97 (1980): 473–79.

Lloyd, Elisabeth. "Language and Ideology in Evolutionary Theory: Reading Cultural Norms into Natural Law." In *Feminism and Science*, ed. Evelyn Fox Keller and Helen Longino. New York: Oxford University Press, 1996.

Spinage, C. A. *A Territorial Antelope: The Uganda Waterbuck.* London: Academic Press, 1982.

Tennent, W.J. "A Note on the Apparent Lowering of the Moral Standards in the Lepidoptera." *Entomologist's Record and Journal of Variation* 99 (1987): 88–9. Cited in Bagemihl, 1999.

Vasey, Paul. "Homosexual Behavior in Primates: A Review of Evidence and Theory" *International Journal of Primatology* 16 (1995): 173–204.

Wolfe, Linda. "Human Evolution and the Sexual Behavior of Female Primates." In *Understanding Behavior: What Primate Studies Tell Us About Human Behavior*, ed., James Loy and Calvin B. Peters. New York: Oxford University Press, 1991.

Tradition and Modernity: Issues in Philosophies of Technological Change

Technology in a Global World

Andrew Feenberg

Introduction

Japan has always been the test case for the universality of Western culture. The Japanese were the first non-Western people to modernize successfully. They built a powerful economy based on Western science and technology. Yet their society remains significantly different from the Western models it imitates. These differences are not merely superficial vestiges of a dying tradition, but show up in the very structure of Japanese science and technology. Is Japan different enough to qualify as an "alternative modernity?" Does it refute or confirm the claims of universalism? These are the questions Japan raises for us today. An early response to these questions comes from Japan itself. In the 1930s the founder of modern Japanese philosophy, Kitaro Nishida, proposed an innovative theory of multicultural modernity. In this chapter, I will consider the Japanese case and introduce Nishida's remarkable theory, one of the first attempts to grasp the philosophical implications of globalization.

I. Two Types of Technological Development

The department store was introduced into Japan in late Meiji by the Mitsui family. They called their store Mitsukoshi. The store was successful and expanded until it was as large as the Western department stores it imitated.[1]

However, in one respect, the Japanese store was quite different from its models: Mitsukoshi had tatami mat floors. This made for some unique problems. Japanese consumers did not usually remove their shoes to enter the small traditional stores in which they were accustomed to shop. Instead, they walked on paving or platforms near the entrance and faced counters behind which salesmen standing on tatami mats hawked their wares. One can still find a few such stores today. Although Mitsukoshi's tatami mat floors were also unsuitable for shoes, customers had to enter the store to shop. And enter they did, sometimes many thousands each day.

At the entrance a check room took charge of customers' shoes and handed them slippers to use on the fragile floors of the store. As the number of customers grew, so did the strain on this system. One day five hundred shoes were misplaced, and the historian of Tokyo, Edward Seidensticker, speculates that this disaster may have slowed acceptance of Western methods of distribution until after the earthquake of 1923 when wooden floors were finally introduced.

This story tells us something we should know by now about technology: it is not merely a means to an end, a neutral tool, but reflects culture, ideology, politics. In this case, two very different nationally specific techniques of flooring came into conflict as an apparently unrelated change occurred in shopping habits. Neither wooden nor tatami mat floors can be considered technically superior, but each does have implications for the understanding of "inside" and "outside" in every area of social life, including, of course, shopping. It eventually became clear at Mitsukoshi that Western methods of distribution required Western floors.

The conflict between these flooring techniques has long since been resolved in favor of Western methods in most public spaces in Japan except traditional restaurants, inns, and temples, where one still removes one's shoes before entering. Nevertheless, the tatami mat conserves a powerful symbolic charge for the Japanese, and many homes have both *washitsu*— Japanese style rooms, and *yoshitsu*—Western style rooms. This duality has come to seem emblematic of Japan's cultural eclecticism. Globalization there has largely meant conserving aspects of traditional Japanese technique, arts and crafts, and customs alongside an ever-growing mass of Western equivalents. At first it seemed that a Western branch had been grafted onto the Japanese tree. Today, one may well ask if it is not a Japanese branch surviving precariously on a tree imported from the West.

This Mitsukoski story illustrates the idea of nationally specific branching development. Branching is a general feature of social and cultural development. Ideas, designs, and customs circulate easily, even among primitive societies, but they are realized in quite different ways as they travel. Although technical development is constrained to some extent by a causal logic, design in this domain, too, is underdetermined, and a variety of possibilities are explored at the inception of any given line of development. Each design corresponds to the interests or vision of a different group of actors.

In some cases the differences are quite considerable, and several competing designs coexist for an extended period. In modern times, however, the market, political regulations, or corporate dominance dictate a decision for one or another design. Once the decision is consolidated, the winning branch is "black-boxed" and placed beyond controversy and question.

It is precisely this last step which did not take place in the relations between national branches of design until quite recently. Poor communications and transport meant that national branches could coexist for centuries, even millennia, without much awareness of each other and without any possibility of decisive victory for one or another design. Globalization is the process of intensified interaction between national branches, leading to conflicts and decisions such as the one illustrated in the Mitsukoshi story.

However, conflict and decision is not the only consequence of a globalized world. Here is a second story that illustrates a different pattern I call "layered" development.[2]

Shortly after the opening of Japan to the world, the Satsuma domain hired a British bandmaster named William Fenton to train the first Japanese military band. Fenton noticed the lack of a Japanese national anthem and set about creating one. He identified a poem, which is still sung as the lyrics of the Japanese national anthem, and set it to music. This unofficial anthem had its debut in 1870, but it was nearly unsingable and quickly fell into disuse.

The need for an anthem was especially pressing in the Navy. Japanese officers were embarrassed by their inability to sing their own anthem at flag ceremonies at sea. The Navy therefore invited court musicians to train the Navy band in traditional Japanese music in hopes that among the performers a composer would be found. But the process was too slow, and the Navy finally asked the court musicians themselves to supply it with suitable compositions. The results were again disappointing. The court musicians came up with a piece in a traditional mode arranged for performance by a traditional ensemble, hardly the sort of thing one would have ready and waiting in a stateroom on a Navy ship!

Around this time, Fenton was replaced by a German bandmaster named Franz Eckert. Herr Eckert rose to the occasion. He arranged the anthem supplied by the court for a Western band, making suitable modifications for playability. In 1880, Japan finally had its current national anthem.

This story is quite different from the Mitsukoshi one. Like flooring, music had developed in Japan and the West along different branches; however, the Japanese national anthem is neither Japanese nor Western, but draws on both traditions. The relations between traditions in this case are quite complex. The very idea of a national anthem is Western. An anthem is a self-affirmation that implies the existence of others before whom the national self is affirmed. But there were no others for Japan during its long 250 years of isolation in a world unto itself. With the opening of the country, self-affirmation became an issue, and an anthem was needed. But how could the anthem affirm Japan unless it reflected Japanese musical style? Hence the composition had to be Japanese. This was easier said than done since the anthem was to be performed by Western instruments at Western-inspired ceremonies. Thus, an original Japanese compositional layer had to be overlaid with a further Western layer in the final stage.

Here we do not have rooms of different styles side by side, but a true synthesis. The merging of traditions takes place in a layering process that is characteristic also of many types of social, cultural, and technological development. Often, several branches can be combined by layering the demands of different actors over a single basic design. In the process, what appeared to be conflicting conceptions turn out to be reconcilable after all. The anthem sounds Japanese played by a brass band. Similarly, modern Japanese politics, literature, painting, architecture, and philosophy emerged in Meiji out of a synthesis of native and Western techniques and visions.

Layering should not be conceived on the model of political compromise, although it does build alliances between groups with initially different, or even hostile, positions. Political compromise involves trade-offs in which each party gives up something to get something. In technological development, as in musical composition, indeed, wherever creative activities have a technical basis of some sort, alliances do not always require trade-offs. Ideally, clever innovations get around obstacles to combining functions and the layered product is better at everything it does, not compromised in its efficiency by trying to do too much. This is what the French philosopher of technology, Gilbert Simondon, calls "concretization."[3] It is this layering process which gives rise to global technology, combining many national achievements in a single fund of world invention.

II. The Globalization of Development

Branching and layering are two fundamental developmental patterns. Their relations change as globalization proceeds. Elsewhere, I have described two styles of design corresponding to different stages in this process. What I call "mediation centered design" characterizes the earlier stage, in which each nation develops its technology relatively independently of the others.[4] Of course, ideas do travel, but the overwhelming weight of particular national traditions ensures that they will be incorporated into devices differently in different contexts. These differences are owing in large part to nationally specific ethical and aesthetic mediations that shape design. Thus, each design "expresses" the national background against which it develops.

Globalization imposes a very different pattern, which I call "system centered design."[5] The globalizing economy develops around an international capital goods market on which each nation finds the elements it requires to construct the technologies it needs. This market moves building blocks such as gears, axles, electric wires, computer chips, and so on. These can be assembled in many different patterns.[6]

The capital goods market is such a tremendous resource that, once interchange between nations intensifies, no one attempts to bypass it. But when design is based on the assembly of prefabricated parts, it can no longer so easily accommodate different national cultures. Instead of expressing a cultural context, products tend more and more to be designed to fit harmoniously into the preexisting system of parts and devices available on the capital goods market. Accommodation to national culture still occurs, of course, but it shares the field with a systematizing imperative that knows no national boundaries. Meanwhile, national culture expresses itself indirectly in the contribution it makes to innovation on the capital goods markets themselves. I would like to develop these two consequences of globalization.

The shift toward system-centered design has implications for the role of valuative mediations in the structure of modern, globalized technology. Traditional technologies generally fit well together. Japanese tatami mat floors, traditional architecture, eating and sleeping habits, shoes, all are of a piece. As such, they express a definite choice of way of life, a valuative framework rooted in Japanese culture. However, on purely technical terms, the links between the artifacts involved are relatively loose. It is true that houses need entryways in which to leave shoes, that futons must be spread on tatami mats, and so on, but adapting each of these artifacts to the others is not very constraining. The wide margin for choice makes it easy for cultural mediations to install themselves in technical design. Indeed, traditional crafts do not distinguish clearly between cultural and technical constraints. There is a "right way" to make things, and it conforms to both.

The globalization of technology changes all this. When design is system based, it must work with very tightly coupled systems of technical elements. Electric wires and sockets cannot be designed independently of the appliances that will use the electricity. Wheels, gears, pulleys, and so on, come in sizes and types fixed by decisions made in their place of origin. A device using them must accommodate the results of those decisions.

System-centered design thus imposes many constraints at an early stage in the design process, constraints that originate in the core countries of the world system. These constraints are imposed on peripheral nations participating in the globalizing process without regard for their national cultures. Furthermore, the very availability of certain types of capital goods reflects the national technological evolution and priorities of the core countries, not those of later recipients. Thus, the effect of globalization is to push cultural constraints to the side, if not to eliminate them altogether. The products that result appear to be culturally "neutral" at first sight, although in fact they still embody cultural assumptions which become evident with wide use in peripheral contexts.

The computer is an obvious example. For us Westerners, the keyboard appears to be technically neutral. But had computers been invented and developed first in Japan, or any other country with an ideographic language, it is unlikely that keyboards would have been selected as an input device for a very long time. Just as the FAX machine prospered first in Japan, so computers would probably have been designed early with graphical or voice inputs of some sort. The arrival of Western computers in Japan was an alienating encounter with the West, a challenge to the national language. Considerable cleverness had to be invested in domesticating the keyboard to Japanese usages.

These observations indicate the weakness of national culture in a globalizing technological system. However, there is another side to the story. Countries far from the core, such as Japan was until quite recently, may not contribute as much as core countries, but they do contribute something. And these contributions will be marked by their national cultural background. In the case of Japan, the magnitude of these contributions has grown to the point where they are a significant factor for the original core countries. Global technology contains a Japanese layer, and so exhibits a true globalizing pattern, not simply core/periphery relations of dependence.

It is difficult to give examples of this feedback from national culture to capital goods' markets. A cultural impulse realized technically looks just like any other technical artifact. Still, a cultural hermeneutics ought to be able to find the cultural traces in the technical domain.

Perhaps miniaturization could be cited as a specific contribution reflecting Japanese culture. At least this is the argument of O-Young Lee, whose book *Smaller Is Better: Japan's Mastery of the Miniature*, argues that the triumph of Japanese microelectronics is rooted in age-old cultural impulses.[7] The impulse to miniaturize, evident in bonzai, haiku poetry, and other aspects of Japanese culture, appears in technical artifacts, too. Lee cites the early case of the folding fan. Flat fans invented in China arrived in Japan very early. The folding fan, which seems to have been invented in Japan in the middle ages, was exported from there to China, inaugurating a familiar pattern. The basic technology of the transistor radio and the videotape recorder both came from the United States, but the miniaturization of the devices, which was essential to their commercial success, took place in Japan, from which they were exported back to the United States.

Of course, once capital goods markets are flooded with miniaturized components, every country in the world can make small products without cultural afterthoughts. But if Lee is right, the origin of this trend would lie in a specific national culture. In a sense, aspects of that culture are communicated worldwide through the technical specifications of its products.

III. Nishida's Theory of the Global World

In the first part of this paper I have illustrated a thesis about the globalization of technology with stories about Japan. In the remainder I will try to draw out the implications of this thesis for the major contribution of Japanese philosophy to the understanding of globalization, Nishida's pre-War theory of the global world.

The context of Nishida's argument was the growing self-assertion of Japan in the early twentieth century. For many Japanese this was primarily a matter of national expansion, but for intellectuals like Nishida, the stakes were still higher—world cultural leadership. These two aspects of Japan's economic and military rise were connected but not identical. On the one hand, Japan had become powerful enough to conquer its neighbors. On the other hand, this very fact showed that Japan, an Asian nation, could participate fully in cultural modernity, assimilating Western achievements, and turning them to its own purpose. Nishida argued on this basis that Asia could finally take its place in the modern world as the cultural equal, or even superior, of the West.[8]

The link between Nishida's position and Japanese imperialism is thus complex and controversial. I have already contributed to that debate in several articles and will return briefly to this topic in the conclusion of this paper.[9] However, my main interest here lies elsewhere, in the parallel I find between the structure of technological globalization as I have explained it above and Nishida's conception of a "global world (*sekaiteki sekai*)."[10] I will show that the contrast between branching and layering underlies this conception, although Nishida misses the technological implications of his approach.

Nishida argues that until modern times, the world had what he calls a "horizontal" structure, that is it consisted of nations lying side by side on a globe that separated, rather than united, them. The concept of "world" was necessarily abstract during the long period that preceded the modern age. By this Nishida means that "world" was a concept only, not an active force in the lives of nations. This condition was unusually prolonged in the case of Japan, which remained disconnected from growing world commerce and communication until the 1860s.

International commerce transformed this horizontal world by bringing all the nations into intense contact with each other. The result was the emergence of what Nishida calls a "vertical" world, a world in which nations struggle for preeminence. Every nation now participates actively in the life of its neighbors—even quite remote nations, through war, trade, and the movement of people and ideas. But there is no harmonious fusion here, but rather a hardening of identities that leads ultimately to war. In this context, nationalism emerges as a survival response to the threat of foreign domination.

Nishida has several other terminologies for this shift that sound rather odd to our contemporary ears, but which are ultimately suggestive. Perhaps the best way to understand his approach is as a dialectic of conceptual frameworks, each one inadequate by itself to describe social reality, but able to do so all together in a mutually correcting system of categories. The complexity of Nishida's argument is supposed, therefore, to correspond to the actual difficulty of thinking global sociality.

Nishida develops the contrast of horizontal and vertical worlds further in terms of the relation of the "many" to the "one" in space and time. The many nations dispersed in space enter into interaction in the modern world. Interaction in history implies more than the mechanical contact of externally related things. Each nation must "express" itself in the world in the sense of enacting the meanings carried in its culture. This can lead to conflict as nations attempt to impose their perspective on all the others. But interaction also requires commonality. Two completely alien entities cannot interact. At each stage in modern history a common framework is supplied by a dominant nation that defines itself as a unifying "world" for all the others. The unification involves the imposition of a general form on the struggle of the particular nations. Nishida gives the example of Great Britain's imposition of the world market on the nineteenth century

(Nishida 1991, 24). The many conflicting nations are thus bound together at a deeper level in one world.

The passage from the many to the one is also reflected in the relations of space and time. The dispersal of the nations in space, their "manyness," is complemented by the simultaneity of their coexistence in a unifying temporal dimension. The struggles of the nations have an outcome that is this unity. Thus in modern times, geography is subordinated to history. The unifying nation represents time for this world and, as such, loses itself in the process of unification it imposes. Britain is absorbed into the world market it creates and becomes the scene on which the world economy operates. The particularity of the nation, Britain, is transcended by the universal order it institutes and for which it stands.

The mechanical and the organic form is yet another terminological couple that Nishida explores. The mechanical world is made of externally related things dispersed in space. Mechanically related things can properly be called individual. Their multiplicity forms an "individual many" (*kobutsuteki ta*) (Nishida 1991, 29–31). The organic world consists of wholes oriented toward a *telos* in time. The whole is thus a subject of action, a "holistic one" (*zentaiteki ichi*) (Nishida 1991, 37–8). Society is not adequately described as mechanical because it forms a whole, and yet it is not organic because its members are fully independent individuals, not a herd. The undecidability of the mechanical and the organic gestures toward the originality of the social world, which cannot be represented by either concept because it embraces both.

Nishida introduces the concept of "place" (*basho*) in a final attempt to conceptualize this "self-contradictory" globalized world. Place, in Nishida's technical sense of the term, is the "third" element or medium "in" which interacting agents meet. Had they nothing in common, they could not meet and interact. But what is it that holds them together? A separate entity would itself require a place to interact with the actors. The *basho* is thus not something external to the interaction but a structure of the interaction itself. This structure arises as each actor "negates itself" to become the "world" for the other, that is, the place of the interaction (Nishida 1991, 30).

It is not easy to interpret this obscure formulation. It seems to mean that in acting, the self becomes an object for the other; it is encountered in the other's path. But the self is not just any object, but the environment to which the other must react in asserting itself as subject. As the other reacts, it defines itself anew, and so its identity depends on the action of the self. But the determination of the other by the self is only half the cycle; the action of the other has an equivalent impact on the self. Interaction is the endless switching of these roles, a circulation of self-transforming realizations (*jikaku*) achieved through contact with an other self.[12]

Nishida has two ways of talking about the role of place in the modern world. Sometimes, he writes as though the globalizing nation serves as the "place" of interaction for the other nations of the world, the scene of interaction. This place can be imposed by domination or freely consented as cultural supremacy, the difference Nishida assumes between England in the past and Japan in the future (Nishida 1991, 99, 77; Nishida 1965c, 373, 349). At other times, he claims that the modern age is about the emergence of global place in the form of a world culture of national encounter.[13] Nishida does not see any contradiction between these two discourses because he assumes that Japanese culture is a kind of "emptiness" capable of welcoming all cultures. But as we will see, this ambiguity turns out to be quite important.

On the basis of this analysis, Nishida asserts the importance of all modern cultures. Western dominance is only a passing phase, about to give way to an age of Asian self-assertion. The destiny of the human race is to fruitfully combine Western and Eastern culture in a "contradictory self-identity." This concept refers to a synthesis of (national) individuality and (global) totality in which the emerging world culture is supposed to consist.

There is a sense in which this global world constitutes a single being, which changes through an inner dynamic. Thus the world "determines itself." But the identities of the particular nations are not lost in this unified object. The resulting world culture will not replace national cultures. Something more subtle is involved. Nishida writes, "A true world culture will be formed only by various cultures preserving their own respective viewpoints, but simultaneously developing themselves through global mediation" (Nishida 1970, 254). World culture is a pure form, a "place" or field of interaction, and not a particularistic alternative to existing national cultures. They persist and are a continuing source of change and progress. The process of self-determination is thus free in the sense of being internally creative; it is not determined by extrinsic forces or atemporal laws. There is nothing "outside" the world that could influence or control it. Even the laws of natural science must be located inside the world as particular historically conditioned acts of thought (Nishida 1991, 36).

Here is a passage in which Nishida describes the global world as he envisages it: "Every nation/people is established on a historical foundation and possesses a world-historical mission, thereby having a historical life of its own. For nations/peoples to form a global world through self-realization and self-transcendence, each must first of all form a particular world *in accordance with its own regional tradition.* These particular worlds, each based on a historical foundation, unite to form a global world. Each nation/people lives its own unique historical life and at the same time joins in a united global world through carrying out a world historical mission" (Nishida 1965a, 428; Arisaka 1996, 101–2).

However, this cosmopolitan argument culminates strangely in the claim that Japan is the center of the unifying tendency of global culture. Just as Britain unified the world through the world market in the spirit of utilitarian individualism, leading to endless competition and strife, so Japan will unify the world around its uniquely accommodating spiritual culture, leading to an age of peace. Japan will be the "place" on which the world will move beyond the limits of the West to become truly global. Japan can lead the world spiritually because its unique culture corresponds to the actual structure of the global world: "It is in discovering the very principles of the self-formation of the contradictory self-identical world at the heart of our historical development that we should offer our contribution to the world. This comes down to practicing the Imperial Way and is the true meaning of 'eight corners under one roof'" (*hakkoo ichiu*) (Nishida 1991, 70).

The vagueness of this conclusion is disturbing. Nishida explicitly condemns imperialism and argues that Japan cannot be the place of world unity if it acts as a "subject" in conflict with other nations. Instead, it must "negate itself" and become the "world" for all other nations (Nishida, 1991, 70, 77). Yet, he also recognizes the fatal inevitability of world conflict and seems to accept Japan's role within that context, as in this statement from his speech to the emperor: "When diverse peoples enter into such a world historical (*sekaishiteki*) relation, there may be conflicts among them such as we see today, but this is only natural. The most world historical (*sekaishiteki*) nation must then serve as a center to stabilize this turbulent period."[14] And, as we see above, he employs ultranationalist slogans with abandon, apparently in the hope of being able to instill new meaning into them. The least that one can say is that his efforts were naive and lent backhanded support to an imperialistic system that conflicted fundamentally with his own philosophical premises.

But just as one can seriously question the depth of the connection between Nazism and Heidegger's thought, if not his actions, similar doubts arise around Nishida's nationalism. There is no clear logical connection between his claims about Japan and his conception of global unity. At least the British gave the world the world market around which to unify. What does Japan have to offer? What mediation does it provide that qualifies it as the center of the new age?

So far as I can tell, Nishida was not bothered by this question, although he should have been. He claims that Japan is the *archetype* of global unity through its ability to assimilate both Eastern and Western culture, but while this is indeed admirable, it is not clear how it qualifies Japan as the *place* of global unity. For that one would think that Japan would have to do something more positive on the world stage than simply to exist as a model. Nishida does announce the world historical significance of the liberation of Asia from Western imperialism. Yet this is certainly not the equivalent of the world market as a unifying force. In the end, this question remains unanswered.[15]

IV. Technology and Place

Despite these problems, I do not think this should be the last word on Nishida's theory of globalization. Once its nationalistic excrescence is removed, the structure of the theory is truly interesting. Nishida's basic claim is that the world has moved from a horizontal to a vertical structure, from indifferent coexistence in space to mutual involvement in time in a conflictual but creative process of global unification. The emerging unity does not efface national differences but incorporates them into an evolving world culture that is best defined as a "place" of encounter and dialogue. A common underlying framework makes possible the communication of nations amidst their conflicts.

This claim precisely parallels the analysis of the passage from branching to layered development presented in the first part of this paper. The various branches of technology in a spatially dispersed world finally meet in the global world of modern times. There they assert themselves and come into conflict, but there they also inform each other with ideas and inventions drawn from diverse national traditions. The outcome, global technology, forms a sort of "place" in Nishida's sense, a scene on which the encounter between nations proceeds with global cultural consequences, but without eliminating the originality and difference of the constitutive national cultures. The layering process in which each culture expresses itself while at the same time contributing to a single fund of invention is thus precisely congruent with Nishida's conception of world culture.

Nishida comes close to making some such connection. He understands that historical action is inextricably intertwined with technical creation. He explains that "Culture includes technique" (Nishida 1991, 61). Technique is an expression of a people's spirit as it interacts with the environment, and through that interaction forms itself (Nishida 1991, 57; Nishida 1965c, 328). "We create things through technique and in creating them we create ourselves" (Nishida 1991, 33; Nishida 1965c, 297). Although Nishida did not do so, one can build on these observations and carry them a step further by relating this social conception of technique to his notion of global cultural interaction in the twentieth century.

Nishida himself was witness to this process as it unfolded in Japan. He was surrounded by rapid social, cultural, and technological change, which he welcomed, and which he believed could become the medium for the expression of an authentic Japanese spirit. He rejected the ultranationalist insistence on keeping the Japanese branch pure in the age of global interaction and insisted that Japan should enter the world scene and move forward. In this he was the theorist of his moment in history, a moment in which Japan appeared to be successfully combining Eastern and Western styles in every domain of life. Nishida lived these events intensely. Perhaps he lost his shoes at Mitsukoshi. Surely, he sang the national anthem and was swept along with his generation by the syncretic modernization of Japan's government, cities, schools, and cultural production. I conjecture that this background underlay his conception of the global world and his confidence in the future. If only he had realized how small a role national politics would ultimately play in that world compared with the force of global technology!

Acknowledgments

I want to thank Yoko Arisaka and Mayuko Uehara for generous help with translations and interpretation of Nishida. They have corrected many misunderstandings; those that remain are my own.

Notes

1. The full account of this story is to be found in Edward Seidensticker, *Low City, High City* (New York: Knopf, 1983).
2. The account below is drawn from William Malm, "The Modern Music of Meiji Japan," in *Tradition and Modernization in Japanese Culture,* ed., Donald Shively (Princeton: Princeton University Press, 1971). For more on layering, see Andrew Feenberg, *Alternative Modernity,* (Los Angeles: University of California Press, 1995a), ch. 9; hereinafter cited in text.
3. Gilbert Simondon, *Du Mode d'Existence des Objets Techniques* (Paris: Aubier, 1958), ch. 1.
4. I formerly called this "expressive design" (Feenberg, 1995a: 225).

5. I formerly called this "system congruent design" (Feenberg, 1995a: 225).

6. For more on the capital goods market, see Nathan Rosenberg, "Economic Development and the Transfer of Technology: Some Historical Perspectives," *Technology and Culture* 11 (1970). Junichi Murata has developed the significance of Rosenberg's analysis for philosophy of technology. See Junichi Murata, "Creativity of Technology and the Modernization Process of Japan," in this volume.

7. O-Young Lee, *Smaller Is Better: Japan's Mastery of the Miniature* (Tokyo: Kodansha, 1984).

8. Kitaro Nishida, *La Culture Japonaise en Question*, trans. Pierre Lavelle (Paris: Publications Orientalistes de France. 1991); hereinafter cited in text.

9. Andrew Feenberg, "The Problem of Modernity in the Philosophy of Nishida," in *Rude Awakenings: Zen, the Kyoto School and the Question of Nationalism*, eds. John Heisig and John Maraldo (Honolulu: University of Hawaii Press, 1995b); Andrew Feenberg, "Experience and Culture: Nishida's Path to the 'Things Themselves'," *Philosophy East and West* 49(1) (January 1999): 28–44.

10. This exposition is based primarily on Nishida 1991.

11. See also Kitaro Nishida, "Nihonbunka no mondai" ("The Problem of Japanese Culture"), *Nishida Kitaro Zenshu* (Tokyo: Iwanami Shoten, 1965c), vol. 12, 291–2, 294; hereinafter cited in text.

12. See, for example, Nishida 1970, 78–79, 134–5; Ohashi, Ryosuke "The World as Group-Theoretical Structure," unpublished manuscript, 1997.

13. Kitaro Nishida, "Sekai Shin Chitsujo no Genri" ("The Principle of New World Order"), in *Nishida Kitaro Zenshu* (Tokyo: Iwanami Shoten, 1965a), vol. 12, 428; hereinafter cited in text. Yoko Arisaka, "The Nishida Enigma," *Monumenta Nipponica* 51 (spring 1996), 101–2; hereinafter cited in text.

14. "Rekishi Tetsugaku ni Tsuite" ("On the Philosophy of History"), in *Nishida Kitaro Zenshu* (Tokyo: Iwanami Shoten, 1965b), vol. 12, 270–1.

15. For an analysis of the debate over Nishida's politics and one of the principal texts under dispute, see Arisaka 1996. For a variety of positions, see John Heisig and John Maraldo, eds., *Rude Awakenings: Zen, the Kyoto School and the Question of Nationalism* (Honolulu: University of Hawaii Press, 1995).

References

Arisaka, Yoko. "The Nishida Enigma." *Monumenta Nipponica* 51 (Spring 1996): 81–105.

Feenberg, Andrew. *Alternative Modernity*. Los Angeles: University of California Press, 1995a.

———. "The Problem of Modernity in the Philosophy of Nishida." In *Rude Awakenings: Zen, the Kyoto School and the Question of Nationalism*, ed. John Heisig and John Maraldo. Honolulu: University of Hawaii Press, 1995b.

———. "Experience and Culture: Nishida's Path to the 'Things Themselves'." *Philosophy East and West* 49(1) (January 1999): 28–44.

Heisig, John, and John Maraldo, eds. *Rude Awakenings: Zen, the Kyoto School and the Question of Nationalism*. Honolulu: University of Hawaii Press, 1995.

Lee, O-Young. *Smaller Is Better: Japan's Mastery of the Miniature.* Tokyo: Kodansha, 1984.

Malm, William. "The Modern Music of Meiji Japan." In *Tradition and Modernization in Japanese Culture,* ed., Donald Shively. Princeton, Princeton University Press, 1971.

Murata, Junichi. "Creativity of Technology and the Modernization Process of Japan," this volume.

Nishida, Kitaro. "Sekai Shin Chitsujo no Genri" ("The Principle of New World Order"). In *Nishida Kitaro Zenshu,* vol. 12. Tokyo: Iwanami Shoten, 1965a.

———. "Rekishi Tetsugaku ni Tsuite" ("On the Philosophy of History"). In *Nishida Kitaro Zenshu,* vol. 12. Tokyo: Iwanami Shoten, 1965b.

———. "Nihonbunka no mondai" ("The Problem of Japanese Culture"). In *Nishida Kitaro Zenshu,* vol. 12. Tokyo: Iwanami Shoten, 1965c.

———. *Fundamental Problems of Philosophy.* Trans. David. Dilworth. Tokyo: Sophia University Press, 1970.

———. *La Culture Japonaise en Question.* Trans. Pierre Lavelle. Paris: Publications Orientalistes de France. 1991.

Ohashi, Ryosuke. "The World as Group-Theoretical Structure." unpublished manuscript, 1997.

Rosenberg, Nathan. "Economic Development and the Transfer of Technology: Some Historical Perspectives." *Technology and Culture* 11 (1970).

Seidensticker, Edward. *Low City, High City.* New York: Knopf, 1983.

Simondon, Gilbert. *Du Mode d'Existence des Objets Techniques.* Paris: Aubier. 1958.

Creativity of Technology and the Modernization Process of Japan

Junichi Murata

Introduction

One of the most conspicuous characteristics of the present situation in technology studies is the dominance of social constructivism in the broadest sense of the word. Many kinds of approaches belong to this trend (cf. Keith Grint and Steve Woolgar 1997, ch. 1). Despite all differences, what makes discussions in these approaches especially interesting is their common stance against the essentialist tendency in some way or other. These approaches emphasize that they do not commit themselves to any determinism, whether technological or social. That means they do not presuppose the naïve distinction between the technical and the social. Rather, they admit that technological development is determined neither by technical nor social factors alone but by each contingent situation in which the sociotechnical network is realized and in which technological artifacts are interpreted correspondingly. Technological artifacts and their ways of working are considered to have no inherent and essential attributes, and their meanings are considered to be open to various interpretations by different social groups. In this sense, "interpretative flexibility" has become a key concept of this trend.

While this nonessentialism is what makes the present discussions of technology studies intriguing, it has engendered several critical views towards this approach. Some see in it a sign of excessive relativism that destroys its own ground, while others criticize it by maintaining that too much flexibility makes a critical stance towards technology impossible (Collins and Yearly 1992; Winner 1991). The difficulty comes to the fore especially when the relationship between modernity and technology is under analysis.

On the one hand, it is difficult to retain a nonessentialist stance towards technology when we consider technology to be one of the essential factors of modernity; it seems that we cannot but assume there is an essential character of modern technology that marks it as different from a traditional one. In fact, we have many well-known conceptual schemes that orient our thinking in an essentialist direction: for example, Heidegger's concept of "Gestell" or M. Horkheimer's concept of "the domination of instrumental rationality." Almost always when such concepts are used to formulate a problem concerning modernity and technology, it is presupposed that modern technology is essentially different from traditional technology. However, when we analyze concrete technological phenomena and search for a concrete criterion for identifying modern technology in contrast to a traditional one, it soon becomes clear that these concepts are too abstract to be helpful.

On the other hand, while researchers belonging to a social contructivist approach concentrate on analyzing how technological artifacts and their ways of working are constituted through a sociotechnical network, they seldom show interest in the problems of how to differentiate modern technology from a premodern one. Perhaps this kind of problem seems to them to be burdened with too many old metaphysical or ideological factors, which presuppose the essentialist way of thinking. We thus find ourselves in a difficult position when we try to deal with the problem concerning the relationship between modernity and technology.

Is there a possible way to deal with the problem of the relationship between technology and modernity without taking an essentialist stance? How can we characterize modern technologies in contrast to traditional ones, while taking their interpretative flexibility seriously?

These are the questions which I would like to discuss in this essay. In the following section, I take up the "creative" character of technology, extending the concept of interpretative flexibility from design and production process to the interactive process between artifacts and users, and also between producers and users. In this discussion I draw on the concepts and perspective of Kitaro Nishida (1870–1945), the preeminent philosopher of modern Japan, whose philosophy could be interpreted as an attempt to develop a nonessentialist way of thinking. In the third section, comparative case studies of technology transfer, mainly in the late nineteenth century and seventeenth century in Japan, are used to illustrate how differently the creative character of technology is realized. In the concluding section, based on stories of modernization in Japan, I suggest a possible answer to the above questions.

"Otherness" and Creativity of Technology

The ambiguous character of technological artifacts.

Instrumentality is one of the most familiar aspects of technology. But if we examine this aspect of technology more closely, its ambiguous character becomes apparent. According to the cognitive theories of artifacts, artifacts are considered to be not only results of intelligent human work, but also the cause of the intelligent behavior of human beings. In order to solve a problem, such as keeping out of the rain, we make an artifact, such as a roof. But once we have made it, we can entrust the work of problem solving to it without bothering again to solve that problem. R. Gregory calls this role of an artifact "potential intelligence" (Gregory 1981, 311ff.).

Gregory puts this role of instrument into a historical order by saying "we are standing on our ancestor's shoulders" (Gregory, 312). When we emphasize the contemporaneous function of the ancestor's accomplishment, which is utilized during the process of problem solving, we could also say that artifacts play a role of "co-actor" in our intelligent and rational behavior.

From this cognitive view we can point out at least two features of artifacts. First, artifacts are made and used by human beings as instruments with which a certain problem can be solved. In this sense an artifact seems to indicate the notion of "means" only because human beings use it for a certain purpose. But secondly, sometimes we are encouraged or compelled to use a specific means for a certain purpose if we want to be intelligent and rational. Artifacts make our intelligent and rational behavior possible. In this way we can already find in the most general characteristic of an instrument an ambiguous feature, which characterizes a means as something more than a simple means.

I would like to call this surplus moment, which is "more than" a simple means, the "otherness" of technology, as it shows a moment that cannot be reduced to a pure instrumental meaning and that sometimes motivates various interpretive activities corresponding to each situation. How can this ambiguous character be made clearer? I think this problem is at the crux of the philosophy of technology because the kind of philosophy of technology we have depends on how we characterize this "otherness" of technology or on which moment of "otherness" of technology we focus.

While an instrumentalist view focuses mainly on the instrumental aspect and pays little attention to the aspect of "otherness," technological determinism emphasizes the aspect of "otherness," in which artifacts appear as if they compel us to use them and determine how to use them. In this sense, instrumentalist and determinist views are generally considered to be opposite positions. However, the "co-actor" role of instruments, which constitutes intelligence and rationality of human behavior, indicates that both views have a common tendency.

For example, in principle it is possible not to use a roof in everyday living. But once a roof is made and widely used, it will be regarded unintelligent, irrational, or even inhuman not to use it. Especially when artifacts are designed to be convenient and easy to use, this way of seeing artifacts becomes unavoidable. But exactly this character of artifacts, that is, the character that artifacts determine the rational way of human action, constitutes the central core of theories embracing technological determinism. In this way, we can pick out a common ground between an instrumentalist or a co-actor view and a determinist view of technology. In either view, once the production process is finished, the artifact becomes a black box no longer susceptible to various interpretations, and both views tend to neglect the interpretative flexibility that can be realized in the interactive process between users and artifacts.

Creativity of technology.

Surely, we are frequently encouraged or even compelled to use particular artifacts in a particular way in order to solve a problem when we want to be rational beings. However, the situation is sometimes far from being well defined and is so ambiguous that there is sometimes an opportunity to develop a new relation between human beings and artifacts.

In fact, in the history of technology, it sometimes happens that invented artifacts bring us a new end-means network in which problems and artifacts are reinterpreted and redefined for purposes far removed from the first intent of the original designer. The Internet is a good example. While its original purpose lay in the military field, it has now created a new form of communication in our everyday life. Automobiles are another example. Before automobiles were invented, produced, and widely used, there was no urgent need to travel faster than a horse-drawn carriage. Only after automobiles have become popular has not being able to travel faster than a horse-drawn carriage become a problem. In this sense new artifacts could be seen not only as problem solvers but also as problem makers.

In addition to these cases, we can find ones in which artifacts are interpreted negatively, contrary to the original intentions of designers. Edward Tenner discusses various cases of this kind. Contrary to the futurist prediction that making paper copies will become unnecessary because of electrical networking, offices are still full of paper. Introducing cheaper security systems in a certain area caused malfunctions and user errors, which decreased the level of security. "Things seemed to be fighting back"(Tenner 1996, ix).

These cases impressively demonstrate the "otherness" of technology, which cannot be reduced either to a simple instrumental nor a deterministic role. This characteristic could be called the creativity of technology, because a new meaning for artifacts is realized, whether the new meaning is interpreted positively or negatively. What is notable in these cases is that the creativity is realized not in the process of design and production but rather in the interactive process between users and artifacts. In order to clarify this creative role of interaction between users and artifacts, I refer to Kitaro Nishida's philosophy.

First, Nishida does not emphasize the familiar instrumental role of technology, with which our life is made convenient and stable; instead, he underscores the role of technology which negates the pregiven structure of our historical world and radically transforms it. Nishida has coined this transformational process "from that which is made to that which makes" (*tsukuraretamono kara tsukurumono e*).

We can interpret this phrase as follows: Nishida emphasizes that the technological process does not end when technological artifacts are produced and handed to users. When the products have left the hands of producers and become independent from them, they have a chance to acquire a new meaning and a new developmental direction through the interaction with users. In this sense, Nishida's view of technology is one in which interpretative flexibility can be found not only in the process of design and production but also in diffusions and uses.

Secondly, Nishida characterizes this creative process with the concept of "reverse determination"(*gyaku gentei*). Perhaps this concept suggests that the creative process is determined by users this time instead of producers. But what Nishida emphasizes is that neither producers nor users alone have a decisive role in determining technological developments. Indeed, a creative process is possible only through the interaction between producers and users, both of whom stand in a contradictory relation. In this sense, Nishida's philosophy of technology can be interpreted as a radical form of nonessentialism.

According to Nishida, a contradictory relation among different factors is a driving force of the historical world that moves from the created to the creating. "Our concrete real world is a world which is a self-contradictory identity of the one and the many and moves from that which is made to that which makes. That means, our concrete real world is a historical world" (Nishida 1949, 110).

Thirdly, Nishida finds this transformational creative structure in various levels of the historical world. Especially in his later years, he tried to characterize the dynamic and critical structure of the world in the twentieth century. In a problematic essay written during World War II, he used the concept of "contradictory identity" to characterize the modern and global structure of the twentieth century world in contrast to the eighteenth century world. According to him, while nations and peoples in the eighteenth century were relatively independent and the concept of the world remained abstract, in the twentieth century, the connections and antagonisms among nations and peoples are so strengthened in a unified world that every nation is forced to transcend itself to its "world historical mission." He calls this character of the modern world "multi-world" (*sekaiteki sekai*) (Nishida 1950, 428). Although this characterization of the modern world remains abstract and problematic because of its political implications, it is certain that Nishida tried to characterize the modernity of the historical world with his idiosyncratic conceptual scheme (Feenberg this volume).

In the next section I will explore the scope of Nishida's theses, taking up concrete historical cases in which the relationship between technology and modernity became a central problem.

Hermeneutics of Technology: Modernization of Japan

Radical transformation.

One of the main characteristics of the modernization of Japan in the late nineteenth century, which began under the military pressure of the Western countries, is that the Japanese quickly understood that, in order to adopt Western modern weapons, it was necessary to introduce various industries, connected with military technology. In order to build and sustain those industries it would also be necessary to adopt the whole Western civilization that formed the background of modern industries.

At the beginning phase of the Meiji era, high officials of the new government, most of whom had experienced the "impact of civilization" during their visit to the Western countries, commonly believed in a deterministic view about the relationship between technology and modernity. They held the view that the engine of the Western modern civilization was industrial technology. In order to modernize Japan, they wished to introduce various advanced technologies as rapidly as possible. In this sense, the rapid speed of various technology transfers was not in response to an urgent demand for them. Rather, there was opposition to hasty introduction, because the social and economic conditions in Japan were insufficient to support them, and the economic results were often disappointing. In fact, the many transplanted technologies such as railways, telegraphs, shipbuilding, and iron manufacturing constituted a program of "industrialization from above," introduced mainly by the initiatives from the Ministry of Engineering.

Even if the process was an "industrialization from above," it did not meet a strong rejection at the grassroots level. Most of the people accepted and even welcomed with enthusiasm the modernization brought by various modern technologies. Although modern transplanted machines such as steam locomotive and railway system did not function successfully in the sense of instrumental rationality, they had a great expressive meaning as a demonstration of Western civilization in the early Meiji era. Tetsurou Nakaoka, historian of technology, describes this characteristic of technology in the following way:

> Enterprises of industrialization in the early Meiji era proved to be not directly useful for the industrialization per se. In a sense they could be considered to be a waste. But what I want to say is that they have played a significant role for the industrialization in reproducing the "impact of civilization" in the mind of people, although this role was indirect. Only when we take this role into consideration can we understand why common people have shown such an extraordinarily active response to industrialization. Through this understanding, we can also come to understand what an important role exhibitions have played in the Meiji era. (Nakaoka 1999, 165)

In fact, during the Meiji era, domestic industrial expositions were held regularly, and when the fifth exposition was held in Osaka in 1903, more than four million people visited it. This fact alone shows how much interest people had in modern technologies. Modern technical artifacts introduced to other sociotechnical networks played not only an instrumental role but also an expressive role, as did machines displayed in exhibitions. A train pulled by steam locomotives could be considered a kind of running show window or advertisement media for modern Western civilization, and people are motivated for a certain interpretative activity and begin to "see" the modern Western world "through" a train.

The character of the radical transformation of society in the process of introducing Western technology in Japan becomes especially conspicuous when we compare it with the case of China. In the late nineteenth century, under similar international conditions, Chinese officials tried to introduce Western military technologies; but they assumed that the West's military technologies could be detached from Western culture, and they tried to defend the Chinese traditional society with Western weapons. In one sense the Chinese response can be considered to be more sober and rational than the Japanese response, as the Chinese interpreted Western technologies as neutral instruments and were not enchanted by them. But the result of their efforts was disastrous for China at least at the end of the nineteenth century. Unexpectedly, China was defeated in the Sino-Japanese War of 1894–95 by a small neighboring country.

What was the main cause of this difference between the Japanese and Chinese responses to Western modernity? Why did the Japanese come to understand the inseparability of technology and modernity, while the Chinese adopted the instrumentalist view of technology?

In order to find an answer to these questions, I go back to another type of encounter a few centuries before this story.

Radical translation.
Early Modern Period in Europe

"The clock, not the steam-engine, is the key machine of the modern industrial age" (Mumford 1934,14). This statement of Lewis Mumford demonstrates that the clock has been recognized as a symbol of the modern machine.

During the change from the medieval to modern period, clocks influenced, and were influenced by, the dynamic change of sociotechnical networks. But what I want to emphasize here is that clocks played a decisive role for the radical change of the metaphysical worldview in Western countries. People did not perceive clocks alone but perceived the world "through" clocks, and the world itself was read as though a clock (Ihde 1990, 61). We could characterize this as a creative role of technology, as it played an important role in transforming Western civilization, but it cannot be reduced to an instrumental role. What is characteristic in the case of clocks is that clocks had no strong social or technological networks in which their creative function could be transferred. Indeed, they were "interpreted" in very different ways. We can clarify this point by contrasting the introduction of Western mechanical clocks to China and Japan, which began in the late sixteenth and early seventeenth centuries.

China

In the seventeenth century, many Christian missionaries from Spain and France visited China to propagate their faith, and in the eighteenth century, many Europeans rushed to China in order to establish commercial ties. They brought various offerings, and mechanical clocks, among them, played an important role, because Chinese emperors showed great interest in such mechanical things and collected many kinds of Western clocks. According to a report by one of the French missionaries in the eighteenth century, in the imperial palace there were more than four thousand clocks which were made by master craftsmen in Paris and London. (Landes 1983, 42) Clocks were displayed together with pictures, porcelains, pottery, and many kinds of playthings in a palace and enjoyed by people belonging to the court. Despite this interest in mechanical things, until the middle of the nineteenth century, clocks were not used as an instrument for time measurement and timekeeping but interpreted mainly as an *objet d' art*.

Japan

In contrast to China, very few Western clocks were imported to Japan during the same period. One reason was that Japan used a temporal hour time system, but more important was that Japanese artisans succeeded in adopting the introduced mechanism to make it move according to the Japanese time system and developed many original types of clocks.

In the temporal hour time system, daytime hours were longer than nighttime hours in summer, and shorter in winter, so that the artisans had to invent complex mechanisms adapting the original mechanisms to correspond to the complexities of the Japanese time system. We could view antique Japanese clocks (*wadokei*) as a successful accomplishment of instrumental rationality, which supports the thesis of social construction of technology. Japanese artisans opened the black box of a Western clock mechanism and redesigned the mechanism to correspond to the world of the Japanese. In this way they showed the interpretative flexibility of technology across different cultures.

We can find three types of interpretations of clocks in these examples. In the first case, clocks were interpreted as something *more* than technical, in the second, something *other* than technical, and in the third, as something *simply* technical. In this sense, the Japanese reaction could be considered to be the most rational and enlightened on technological grounds in the narrow sense of the word.

On the other hand, in the case of Japan, we have seen two types of interpretation which make a clear contrast. In the first type, which we have seen in the modernization process in the late nineteenth century, a radical transformation of a sociotechnical network occurred, and the technological determinism was the result of the interpretative activities of Japanese people. In the second type, which we have seen in the transfer process of mechanical clocks in the sixteenth and seventeenth centuries, a radical translation of new artifacts (Western clocks) into a traditional network occurred, and the instrumental rationality realized by Japanese artisans functioned efficiently.

Why did the Japanese show such an enthusiasm for Western technologies in the late nineteenth century, while they were so "rational" about Western clocks earlier? In other words, why did the surplus moment embodied in modern machines in the late nineteenth century not remain in the ideological dimension but in fact have a material influence in Japanese society? Why weren't modern machines detached from their (Western) sociotechnical network, as in the case of the clocks in the seventeenth or eighteenth century?

In the last section, I will first try to pick out an answer to these questions, focusing on the role of artisans, which illuminates an interesting relation between the two types of the technological transfer in the case of Japan. Secondly, based on the results of these discussions, I will try to formulate an answer to the original question concerning the relationship between technology and modernity in general.

Lessons from the Japanese Experience

The role of artisans and the capital goods industry.

The artisans who developed and produced a Japanese style of clock were closely connected to another innovation in the Tokugawa Edo period, the most intriguing of all innovations, the automaton (*karakuri*).

One of the most famous artisans in this tradition of technology was Hisashige Tanaka (1799–1881), who built a very impressive astronomical instrument in the Edo period. In the last days of Tokugawa Edo period, Tanaka was invited by Saga Domain to advise on the technological modernization of steam engines of ships and of guns, among other things. In 1875 he established a private machine-making firm, which later became part of the twentieth-century manufacturing giant Toshiba (Morris-Suzuki 1994, 53).

The path from Japanese clock through automaton to advanced technology was not direct, but rather complicated. The gap between traditional technology and most advanced Western technology was huge at the time. In the iron industry or railroad industry, for example, almost every part of the machines had to be imported, at least at the first phase of their transportation. On the other hand, most of the transported technologies could not be easily rooted in the new context, and only after the hard work of making a new network of various social and technological factors by Japanese engineers and artisans could they function successfully. Without such work of translation, there could not occur a radical transformation of the network. In this context, the capital goods industry and traditional artisans played significant roles in transforming an old network and making a new one.

According to Rosenberg, a capital goods industry plays an important role in the development and transfer of a technology by preparing an appropriate environment for repair and maintenance and successful performance of machines. In order to show these characteristics he describes the innovative role of the machine tool industry in Western Europe in the nineteenth century. Rosenberg also emphasizes the character of technology of the capital goods industry, which is not explicitly codified but is incorporated in skilled personnel, so that there was sometimes a need for "migration" of artisans to realize a transfer of the technological "know-how" that they had.

It is exactly this role of the capital goods industry, indicated by Rosenberg, that corresponds to the role of traditional artisans in the modernization process in Japan. In this way, we can find a "material" background for the rapid introduction of many kinds of Western technologies in late nineteenth century Japan and a technological basis of the enthusiastic response of the people at that time.

Characteristics of modern technology are sometimes considered to be universal and context-independent, in contrast to traditional technology, which is considered to be embedded in a local cultural context. However, if what we have seen here is correct, without an environment provided by traditional technologies, modern technologies cannot be transferred and introduced into other contexts. In this sense, we could say it is rather the developmental processes, translation and transformation processes of traditional technology, that make the modernity of technology possible. Without such support from traditional technologies, the ideological character of modern technology could not be transformed into reality. Modernity, without the help of tradition, would remain only an ideology.

Technology and modernity.

What kind of general answer can we have now to our original question? How can we characterize modern technology in contrast to a traditional one, while taking its interpretative flexibility seriously?

One of the most conspicuous characteristics of the modernization process in Japan is the dual structure of its sociotechnical network with an advanced sector of modern technology and a parallel domestic sector of traditional technology. The advanced sector functions as if transferred technology guides and determines the way of modernization. In reality, however, the advanced sector interacts with the domestic sector, where traditional technology plays a role of instrumental rationality, decreasing the gap between the two sectors sufficiently that advanced technology becomes adapted to local practices. Through this interaction, the scope of flexibility is restricted, the process is channeled in a certain direction, and rapid and continuous adaptation and development of technology becomes possible.

What made the modernization process of Japan possible were these seemingly contradictory, yet inseparably connected, technology sectors. If these factors had been too contradictory, there would have been no successful process of modernization, as was the case during an encounter between China and Western civilization in the late nineteenth century. The gap between the two sectors was too large to be filled at the time in China. On the other hand, if the two sectors had not been contradictory enough, there would have been no radical transformation, as happened in the encounter between Japanese artisans and Western clocks in the early seventeenth century. In this sense, the manner in which the creativity of technology is realized depends on each historical context; it is thoroughly contingent, and we cannot generalize the lessons of the Japanese experience. What we can say is that modernity does not exist in a universal sense, but in modernity there is always a dual structure constituting modern factors and traditional factors. In this sense there are always various modernities (plural) together with various transformational processes of tradition.

On the other hand, we have already figured out a relatively general and formal structure of modern technology, in which capital goods industry plays a decisive role. The creative process can be found in any process of technological development since the beginning of the history of human technology. But what is characteristic in the modern age is that this process is not a random phenomenon but is institutionalized in a sociotechnical network that has a particular dynamic in which technologies are continually transformed.

Since the latter half of the nineteenth century, the international connections between different countries and different cultures have strengthened, and the global character of the world has begun to become conspicuous. While capital goods industries support this global tendency by accelerating the interactions between producers and users in various fields, they are also supported and oriented by this tendency (Feenberg this volume). Different and heterogeneous parts in the sociotechnical network of the modern world are not indifferent to each other and are always involved in a contradictory interactive process that occurs between them. In this way, the interaction between producers and users does not remain stable, but rather is always involved in a transformational process, where, in the words of Nishida, the "reverse determination" leads to conspicuously creative results.

References

This chapter is a short version of a paper which will be published under the title "Creativity of Technology: an Origin of Modernity?" in *Technology and Modernity*, ed., Tom Misa, Philip Brey, and Andrew Feenberg, Cambridge, Mass.: MIT Press, 2002.

Collins, Harry, and Steven Yearly. "Epistemological Chicken." In *Science as Practice and Culture*. ed. Andrew Pickering. Chicago: Chicago University Press, 1992.

Feenberg, Andrew. "Technology in a Global World", in this volume.

Gregory, Richard. *Mind in Science*. London: Penguin Books, 1981.

Grint, Keith, and Steve Woolgar. *The Machine at Work*. Cambridge, U.K.: Polity Press, 1997.

Hacker, Barton. "The Weapons of the West: Military Technology and Modernization in 19th-Century China and Japan." *Technology and the West: A Historical Anthology from Technology and Culture*, ed. T. S. Reynolds and S. H. Cutcliffe. Chicago: University of Chicago Press, 1997.

Ihde, Don. *Technology and the Lifeworld*. Bloomington, Ind.: Indiana University Press, 1990.

Landes, David. *Revolution in Time. Clocks and the Making of the Modern World*. Cambridge, Mass.: Harvard University Press, 1983.

Morris-Suzuki, Tessa. *The Technological Transformation of Japan, From the Seventeenth to the Twenty-first Century*. Cambridge, U.K.: Cambridge University Press, 1994.

Mumford, Lewis. *Technics and Civilization*. London: Routledge, 1934.

Nakaoka, Tetsuro. *Jidousha ga hashitta* (The cars have run). Tokyo: Asahi-Shinbun-sha, 1999.

Nishida, Kitaro. *Complete Edition* Vol. 9, Tokyo:Iwanami-shoten, 1949.

———. *Complete Edition* Vol.12, Tokyo: Iwanami-shoten,1950.

Norman, Donald. *Things that Make Us Smart; Defending Human Attributes in the Age of the Machine.* Reading, Mass.: Addison-Wesley Publishing Company, 1993.

Rosenberg, Nathan. "Economic Development and the Transfer of Technology: Some Historical Perspectives." *Technology and Culture* 11, 1970: 550–575.

Tenner, Edward. *Why Things Bite Back: Technology and the Revenge Effect.* London: Fourth Estate, 1996.

Winner, Langdon. "Upon Opening the Black Box and Finding It Empty: Social Constructivism and the Philosophy of Technology." *Science, Technology and Human Values* 18(3) 1993: 362–378.

Contributors

Licia Carlson is Assistant Professor of Philosophy at Seattle University. Her research interests include feminist philosophy, bioethics, philosophy of disability, twentieth-century French philosophy, and the philosophy of music. She has published articles on Michel Foucault, feminism and disability, prenatal testing, and she is currently completing a book on philosophy and cognitive disability.

Robert P. Crease is Professor of Philosophy at SUNY, Stony Brook, and historian at Brookhaven National Laboratory. His books include *Making Physics: A Biography of Brookhaven National Laboratory* and *The Play of Nature: Experimentation as Performance*. He also writes a monthly column, "Critical Point" for *Physics World*.

Margaret Cuonzo is Assistant Professor of Philosophy at Long Island University's Brooklyn Campus, where she teaches logic, philosophy of science, philosophy of language, and feminist philosophy of science. She is the author of two essays on philosophical paradoxes, "Why the Paradox Has a Restricted Solution at Best," (*Facta Philosophica*) and "Intuition and Paradox" (*The Logica Yearbook 2000*). Cuonzo has presented papers on the philosophy of language and science at the American Philosophical Association (APA), the American Association of Philosophy Teachers (AAPT), the Czech Academy of Sciences, and various universities.

Andrew Feenberg is Professor of Philosophy at San Diego State University. He is the author of *Transforming Technology, When Poetry Ruled the Streets, Questioning Technology, Alternative Modernity, Lukács, Marx and the Sources of Critical Theory,* and is co-editor of *Technology and the Politics of Knowledge*. He has pursued his studies of the philosophy of technology and online education with support from the National Science Foundation, the U.S. Department of Education, and the Digital Equipment Corporation.

Robert Figueroa is visiting Assistant Professor of Philosophy at Colgate University. He is interested in bringing philosophical contributions to the environmental justice movement. His contributions to this effort are published in *The Blackwell Companion to Environmental Philosophy; The Encyclopedia of Religion and Nature;* the second edition of *Faces of Environmental Racism: Confronting Issues of Global Justice,* and the *Environmental Justice Reader: Politics, Poetry, and Pedagogy.* He is currently working on issues pertaining to cultural identity, epistemology, and science and technology studies, especially as these pertain to environmental justice and ethics.

Sara Goering is Assistant Professor of Philosophy and Director of the Center for Applied Ethics in the Department of Philosophy, California State University, Long Beach. Her work in bioethics covers genetic engineering, disability, race, and feminist theory. She is co-editor, with Annette Dula, of *"It Just Ain't Fair": The Ethics of Health Care for African Americans.* She is currently working on an interdisciplinary survey study on ethics, cosmetic surgery, and norms of appearance.

Sandra Harding teaches in the Graduate School of Education and Information Studies and the Women's Studies Program at the University of California at Los Angeles. She co-edits *Signs: Journal of Women in Culture and Society.* She is the editor or author of twelve books on feminist and postcolonial epistemology, philosophy of science, and methodology. These include *Discovering Reality: Feminist Perspectives on Epistemology, Metaphysics, Methodology and Philosophy of Science; The Science Question in Feminism; Feminism and Methodology; Whose Science? Whose Knowledge?; Is Science Multicultural? Postcolonialisms, Feminisms and Epistemologies,* and a forthcoming edited collection, *The Standpoint Reader.*

Robert Hood's research concerns intersections between biomedical and environmental ethics. Currently, he is developing a clinical approach to ecosystem health and environmental ethics. Hood is Assistant Professor of Philosophy at Middle Tennessee State University and Clinical Ethics Fellow at St. Thomas Hospital / Ascension Health in Nashville, Tennessee.

Hugh Lacey is Scheuer Family Professor of Humanities and Professor of Philosophy at Swarthmore College. He is also frequently a visiting professor at University of São Paulo (Brazil). His recent publications include *Valores e atividade cientifica* (São Paulo: Discurso Editorial, 1998), *Is science value free? Values and scientific understanding* (London: Routledge, 1999), and several articles exploring issues of both ethics and the philosophy of science connected with current controversies on genetically engineered (transgenic) crops and alternative forms of agriculture.

James Maffie is Assistant Professor of Philosophy, Colorado State University. He has authored articles on naturalized epistemology, evolutionary epistemology, comparative epistemology, and Conquest-era Nahua philosophy, and recently edited a special issue of *Social Epistemology* devoted to the topic of truth from the perspective of comparative world philosophy. He is currently studying Nahuatl and writing a book entitled, *'Flower and Song' in 'The House of Paintings': Nahua Epistemology in the Era of the Conquest.*

Junichi Murata studied at the University of Tokyo, was a lecturer and an Associate Professor at Toyo University, and is now Professor at the University of Tokyo, Department of History and Philosophy of Science. His publications include *Perception and the Life-World* (in Japanese), University of Tokyo Press, 1995; and "Consciousness and the mind-body problem" in *Cognition, Computation, Consciousness*, ed. M. Ito et al., Oxford University Press, 1997.

Anita Silvers, Professor of Philosophy at San Francisco State University, has published seven books and more than a hundred book chapters and articles, including *Medicine and Social Justice* (with Rosamond Rhodes and Margaret Battin); *Americans With Disabilities: Exploring Implications of the Law for Individuals and Institutions* (with Leslie Francis); *Disability, Difference, Discrimination: Perspectives on Justice in Bioethics and Public Policy* (with David Wasserman and Mary Mahowald); *Sociobiology and Human Nature* (with Michael Gregory); and *The Recombinant DNA Controversy* (with Michael Gregory). In 2002, she co-directed (with Eva Kittay) an NEH Summer Seminar on "Justice, Equality, and the Challenge of Disability."

Michael Ashley Stein, JD, Ph.D., is Associate Professor at the School of Law of the College of William and Mary, where he teaches, among other subjects, a course on Disability Law. In 2001–2002, he served as a Merit Research Fellow for the National Institute on Disability and Rehabilitation Research. He is a member of the advisory boards of several disability rights organizations, including the recently founded *Autonomy*.

Sara Waller received her Ph.D. in philosophy at Loyola University Chicago in 1999. She was trained in psychometric testing at the U.C.S.D. Pediatric Neurology Laboratory. She currently teaches philosophy at California State University, Dominguez Hills.

Alison Wylie is Professor of Philosophy at Washington University-St. Louis. She is centrally interested in questions about standards of evidence and ideals of objectivity in the social sciences (archeology) and in feminist philosophy of science. She is the author of *Thinking From Things: Essays in the Philosophy of Archaeology* (2002), and the coeditor of *Critical Traditions in Contemporary Archaeology: Essays in the History, Philosophy, and Socio-Politics of Archaeology* (1989), *Ethics in American Archaeology* (2000), and *Equity Issues for Women in Archaeology* (1994).

Naomi Zack is Professor of Philosophy at the University of Oregon, Eugene. Her most recent book is *Philosophy of Science and Race* (Routledge 2002), and she is also the author of *Race and Mixed Race; Bachelors of Science: Seventeenth Century Identity, Then and Now,* and a textbook, *Thinking About Race.* She is the editor of *Women of Color and Philosophy* and other anthologies and has written articles on race, racism, mixed race, gender and seventeenth-century philosophy.

Index